THE AGRARIAN SYSTEM OF EASTERN RAJASTHAN (c.1650—c.1750)

THE AGRARIAN SYSTEM OF EASTERN RAJASTHAN

(c.1650—c.1750)

SATYA PRAKASH GUPTA

CENTRE OF ADVANCED STUDY IN HISTORY
Aligarh Muslim University

MANOHAR
DELHI

ISBN 81-85054-11-8

First Published 1986

Published by

Ramesh Jain
for Manohar Publications
1 Ansari Road, Daryaganj
Delhi-110002

Printed by

Lohia Composing Agency
at Sunil Printers
CB-1067, 75/1, Ajay Palace, Naraina
New Delhi-110028

To the Sacred memory of
my teacher
Mrs. Khursheed Nurul Hasan

Contents

Preface ix
Introduction xi
Abbreviations xv

CHAPTERS

I The Amber Principality—Its Nature and Territories 1
II Agricultural Production 38
III Agricultural Prices 74
IV Agrarian Trade 92
V The Structure of Land Rights and the Agrarian Society 116
VI Superior Rural Classes 134
VII The System of Rural Taxation 144
VIII Land Revenue Administration 163
IX The Jāgīr System during the Evolution of the Jaipur State 211
X The Farming System 223

APPENDIX (A)

A-1 Raqbo Pargana Mauzabad 1641 A.D. 233
A-2 Jama' Bandi Mauza Jo Singh Pur Pargana Malarna, 1664 A.D. 244
A-3 Yaddashti Haqiqati Hal Bail Zubāni Paṭel Paṭwāri Pargana Chatsu, 1666 A.D. 255
A-4 Dastūr-'Amal and 'Amal-Dastūr Chabutra Kotwāli, qasba Phagui 1713 A.D. 269
A-5 Dastūr-ul 'Amal Pargana Uda , 1715 A.D. 280
A-6 Taqsīm Pandrahsala Pargana Antela Bhabhra, 1649-1663 A.D. 289

Taqsīm Dahsala, Pargana Udai, (1734-1743A.D.) 289
A-7 K͟hasra of Village Morara, Pargana Naraina, 1745 A.D. 294

APPENDIX (B)

B-1 Silver Value of Gold 310
B-2 Silver Value of Copper 313
Bibliography 317
Additions and Corrections 339
Index 341

Maps
(1) The Parganas Held in *Jagīr* by the Rules of Amber c.1740 *facing page 1*.
(2) Six Central *Parganas facing page 37*.
Figures :
(1) Prices of Major Crops (Figures 1-7) 76-82
(2) Weighted Price and Rate Indices (Figure 8) 87
(3) Running Decennial Averages of Weighted Price Indices of Copper Prices (Figure 9) 90
(4) Indices of Copper Prices (Figure 10) 91
(5) Relative size of Assessments in kind and cash (Figures 1-6) 94-99

Preface

This book is a revised version of my Ph.D. thesis submitted at the Aligarh Muslim University in 1974. In the course of revision some fresh evidence has been examined, and some new problems such as those of agrarian trade, the structure of village society and superior right-holders have been studied in detail.

The present study was undertaken on the advice of my teacher, the late Mrs. Khursheed Nurul Hasan. She first drew my attention to the importance of the Rajasthani records and collaborated with Professor S. Nurul Hasan and myself in an early paper on the pattern of agricultural production in the Amber State.

I owe a great debt of gratitude to Professor S. Nurul Hasan for the critical evaluation of my arguments and conclusions, as also for guidance in the handling of statistical information.

Professor Satish Chandra, who was the first Supervisor of my thesis, took keen interest in my work at all stages, even after he had left Aligarh. I am most beholden to him for a number of insights and other help.

I am deeply grateful to Professor Irfan Habib for his ungrudging interest in the preparation of the book. He was gracious enough to go through the manuscript of the book.

I have received help and encouragement from my colleagues and friends in the preparation of the thesis. Among others, I would particularly like to thank Dr Rajkumar Trivedi and Dr Shireen Moosvi. Dr. Moosvi helped me in the computation of statistical data and putting diacritical marks on Persian words. My student, Miss Sumbul Halim, has compared the figures in my tables.

My thanks are due to Messrs Zahoor Ali Khan and Faiz Habib for the maps and charts they have drawn for this book. Mr M.A. Khan Afridi prepared the typescript.

I am also thankful to the staff of the Rajasthan State Archives, Jaipur and Bikaner, and of the Research Library of the Centre of Advanced Study, Aligarh Muslim University, for the help constantly extended to me.

Finally, I must acknowledge my deep gratitude to my wife, Sneh Gupta, for her indulgence in many ways during the preparation of this book.

SATYA PRAKASH GUPTA

Introduction

The agrarian system of India during Mughal times has been studied on a grand scale, eg., by Moreland (1929), Irfan Habib (1963), Noman Ahmad Siddiqi (1970), and S. Nurul Hasan (1973). These works examine the agrarian structure of India, or at least of Northern India, under the assumption that it was, in most respects, uniform, and the regional differences, if any, can be taken as exceptions that merely prove the rule. It is not that such an assumption is incorrect; indeed, it will be seen that the present writer has much sympathy with it. But it is necessary to test it by studying the agrarian conditions of particular regions in rigorous isolation, and then comparing it with the standard pattern suggested for the Mughal Empire as a whole. Only then would one be able to say with any degree of certainty whether the general assumption has full validity.

In carrying out a regional study of this kind, one is hampered in most cases by a lack of detailed information specific to that region. One would be tempted to accept the tenor of the Mughal administrative decrees and documents, seemingly applicable to the whole Empire, as true for the particular region as well. It will be immediately seen that such a course would 'contaminate' our evidence : we too would be basing ourselves on the same assumption which we are setting out to test.

If there is one region where the strictly local evidence for the Mughal period is so massive as to enable us to construct a picture on that evidence alone, it is Rajasthan, and, within it, especially Eastern Rajasthan. For Marwar, we have Nainsi's celebrated *Vigat*, but it is for the areas in Eastern Rajasthan which passed under the authority of the rulers of Amber that the primary documents have come down to us in the most splendid profusion. For the archives of the Amber/Jaipur administration from about the mid-17th century have survived to a degree unparalleled in Northern India. These are now preserved, practically in their entirety, in the Rajasthan State Archives, Bikaner.

If these archives sometimes appear to be incomplete, this is mainly because of the nature of the administrative authority of the rulers of Amber. As will be seen in Chapter I, the rulers held various *parganas* in the vicinity of Amber as *jāgīrdārs* ; and, by the very nature of the Mughal *jāgīr* system, the *jāgīrs* were transferable so that the records of a *pargana* would be inducted into the archives when it was held in *jāgīr* ; but there would be no records relating to it when it was placed in some other noble's *jāgīr*. The area came to be more consolidated during the first half of the 18th century, when Jai Singh Sawai (1688-1743) gradually took over much of the area under permanent *jāgīr* or on farm.

Keeping in view the relative completeness of the documentary evidence about them. I have mainly scrutinised the records of six contiguous *parganas*, which remained under the authority of the Amber rulers most consistently : Amber (later Jaipur), Bahatri (Baswa), Chatsu, Dausa, Lalsot and Malarna (See Map 2). No hesitation, however, has been shown in utilizing the evidence of other *parganas* comprising Jaipur State, where it was felt to be necessary. The period for which the evidence is studied is c. 1650-1750, since it is for these hundred years, and especially for the last of these, that the evidence is the richest. Owing to Maratha inroads the documentary evidence becomes defective after 1750 ; but in any case, the intention of this work is to confine focus on what can be legitimately designated the Mughal period, which closes by about 1750.

Our documentary evidence is not only large in volume but varied in nature. Roughly, these may be categorised as follows :

A. Documents giving various village-level details, area, tax on individuals, etc., such as *khasra, raqba-bandī, taqsim, terīj, yāddāshtī* of *hal* and *baīl* or *asāmī* and *bail, yāddāshtī haqīqatī zubānī paṭel* and *paṭwārī* etc.

B. Documents containing accounts of items of income, expenditure and balances at the village and *pargana*-level, such as *arhsatta, jama'-bandī, jāma'-kharch, muwāzana*, etc.

C. Price-lists, *nirkhbāzār*, *nirkh-roznāma*, etc.
D. Collections of letters and documents issued by officials such as *chiṭṭhīs*, *parwānas*, *sanads*, *'arẓdāshts*, etc.
E. Administrative manuals such as *dastūr*, *dastūr-ul 'amal 'amal-dastūr*, etc.
F. Correspondence of the Amber/Jaipur court such as *Vakīl's* reports, *khuṭūṭ-i-Mahārājgān* (dated an undated) etc.

These documents will be commented upon in Appendix A and in the Bibliography.

A few points may however be made here regarding the use of these documents. They are, in their bulk, in Rajasthani and are mostly dated in *Samvat* era though the earlier documents often use the *Hijari* era, and the *Faṣlī* calendar is also used. For convenience I have uniformly converted the *Samvat* year into the Christian year by deducting 57 years from it. Since this is done uniformly all the dates remain comparable. In most cases I have also reproduced the full *Samvat* date as given in the original (*Vaḍī*, *Sudī*, date, month and year).

There is a small problem relating to transliteration. A very large number of Persian terms are used in the Rajasthani documents and the question arises whether one should restore their original spelling in transliteration or simply transliterate them as they are spelt in Rajasthani. Are we to write *naqd* (Persian) or *nakad* or *naqad* (Rajasthani) ; *la'iq-i-zirā'at* (Persian) or *layak jarati* (Rajasthani). I have tried to make a compromise by first giving the Rajasthani form and its Persian equivalent, and then, in most cases, giving the Persian form in transliteration, with some rare exceptions. Diacritical marks have been put on the Persian terms, but not generally on the Hindi or Rajasthani words and names.

The evidence that one gets enables one to cover a very large area of what can be properly brought under the term 'agrarian system' : the area under cultivation, the crops, the social divisions (castes) among peasants, the incidence of land-tax, the structure of rights, the framework of revenue administration, the organisation of rural trade, and so on. Undoubtedly, the light cast is rather uneven, wonderfully and unexpectedly sharp on one point, very blurred and indistinct on another. Without pretending that we know as much as we should like to on all the themes that I have just specified, I have attempted a description (and hopefully, a critical analysis) based strictly on the original contemporary material. I hope that the picture that emerges is internally consistent, if not entirely comprehensive.

The reader will see for himself what conclusions emerge from my study : whether it confirms or contradicts the larger analyses of the Mughal agrarian system, and whether what Eastern Rajasthan possessed was in large part unique. But I may anticipate conclusions of the study by saying that the case for uniqueness is little supported by the evidence which is laid out here. Professor Satish Chandra was probably a pioneer in rejecting the case for uniqueness, and now that I have examined the evidence in detail, I can only acknowledge the acuteness of that insight.

Abbreviations

Ā'īn	*Ā'īn-i Akbari*
A.N.	*Akbar Nāma*
D.H.	*Daftar Ḥuzūrī*, Jaipur (now at Bikaner)
IESHR	*The Indian Economic and Social History Review*, New Delhi.
IHR	*Indian Historical Review*, New Delhi.
JASB or *JRASB*	*Journal of (Royal) Asiatic Society of Bengal*, Calcutta.
JESHO	*Journal of the Economic and Social History of the Orient*, Leiden.
J.	*Jāgīr*
JRAS	*Journal of the Royal Asiatic Society*, London.
OHR	*Old Historical Records*, Jaipur (now at Bikaner).
PIHC	*Proceedings of the Indian History Congress.*
PRHC	*Proceedings of the Rajasthan History Congress.*
RSA	Rajasthan State Archives, Bikaner.
Vigat	*Marwar-ra-Pargana-ri-Vigat*
VS	*Vikrama Samvat*
Vill.	Village

MAP-1

EASTERN RAJASTHAN

Parganas held in Jagir by the rulers of Amber, c. 1740

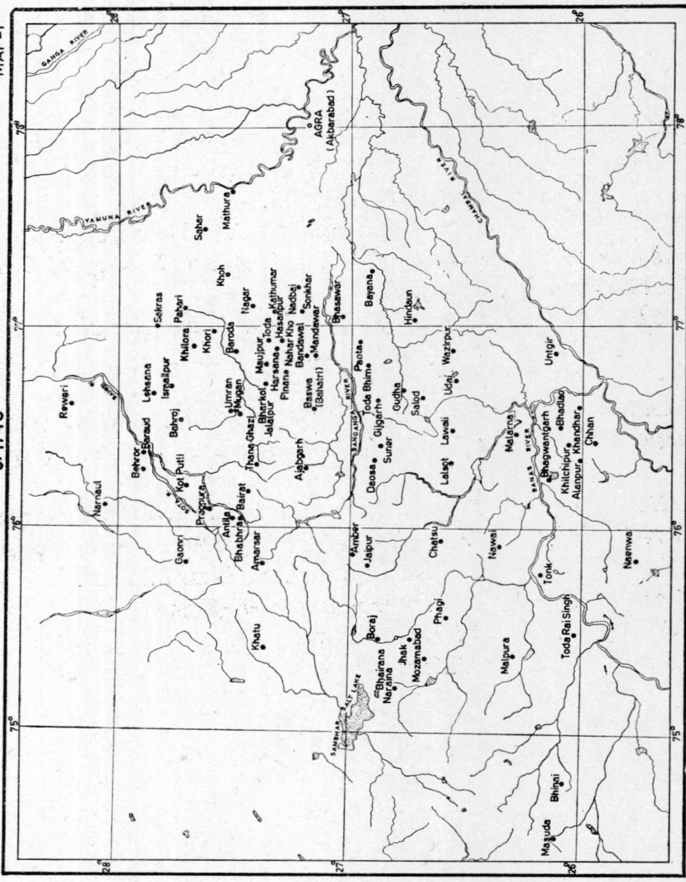

CHAPTER I

The Amber Principality— Its Nature and Territories

I

The Position of Kachhwaha Rulers in the Mughal Empire

The present study is concerned with agrarian conditions in those areas of Eastern Rajasthan which lay under the control of the Kachhwaha rulers of Amber (later of Sawai Jaipur). In time, these came to comprise the bulk of Eastern Rajasthan, as the Kachhwaha principality rose to a premier position among the principalities of that region.

This development took place almost entirely as a result of the close association that the Kachhwahas had with the Mughal Empire. Before the time of Akbar, the domains of the Kachhwahas had been divided up between their two major clans, the Rajavats and Shekhavats; and the rulers of Amber themselves were hard-pressed by their Rajput neighbours and then by the Mughal commandant of Mewat, Mirza Sharfuddin, who imposed harsh terms upon the Kachhwaha ruler, Bharamal in 1562. But Bharamal met Akbar at Sanganer (near Jaipur), as he was going to Ajmer the same year; and the foundations were laid of the most fruitful alliance between the Mughal emperor and the Kachhwaha chief.

The terms of this alliance protected and enriched the Amber rulers, while also altering their position and status. The hereditary Kachhwaha chiefs of Amber had used the titles of Rāja and Mahārāja for themselves. But they were designated *zamīndārs* by the Mughal Chancery, and their ancestral domains were treated as their *waṭan*. To the Mughal court, even the title Rāja (in its Persian spelling) was dependent upon a positive grant of it to the Amber chief by the Mug͟hal Emperor. After Bharamal's submission to Akbar in 1562 the Kachhwaha ruler was formally enrolled as an officer of the Empire, the position of the ruler rising from that of a chief or a *zamīndār* to a member of the Mug͟hal nobility. The celebrated Rajput policy of Akbar was in fact initiated with the arrangement he made with Bharamal. The latter in the words of Abu'l Faẓl, "left the circle of *zamīndārs* and became one of the select of the court."[1] The policy consisted, in essence, of treating the Rajput rulers as nobles at par with others obliged to render military services while receiving *jāgīrs* in lieu of salary.

Territories recognised as that of the chiefs by hereditary right were regarded as their '*waṭan*' *jāgīr*. This meant that, theoretically, the chiefs were entitled to the full revenues of their domains with no obligation to pay tribute. The *waṭan jāgīr* was assigned at a notional revenue-figure (*jama*'), which was appropriated towards meeting the salary of the chief and pay-claims of the contingents (*tābīnān*) kept by him for imperial use. If his status was raised and the required contingent enlarged, by increasing his *zāt* and *sawār manṣabs*, the additional pay-claims were met by assignment to him of other territories in ordinary transferable *jāgīrs*, the so-called *tank͟hwāh jāgīrs*. The revenue derived from the latter *jāgīr* was sometimes much more than that of the chieftains' hereditary domains; and this was almost always the case with the Amber rulers who normally obtained very high *manṣabs*, while their own hereditary domains were relatively small; they thus became entitled to large *jāgīrs* outside their *waṭan*.

The Mug͟hals consistently asserted the principle of 'paramountcy' in their relations with the Rajput chiefs.[2] This meant that a chieftain depended for his position on the goodwill of the emperor as well as on his own inherent rights. Bahadur Shah resumed Amber to the *k͟hāliṣa* (territory under the emperor's own revenue administration) in 1708, since Jai Singh Sawai had sided with Azam during the war of succession. Amber was

1. *Akbarnāma*, II, Bib. Ind. 1897-1921, p. 63.
2. S. Nurul Hasan, '*Zamīndārs* under the Mug͟hals,' *Land Control and Social Structure in Indian History*, ed. R.E. Frykenberg, London, 1969, pp. 17-31.

indeed garrisoned under the Imperial *faujdār*, Saiyad Husain Khan, for a few months.[3] The Mughals also asserted their right of paramountcy in the case of Jodhpur more than once. While generally conforming to the law of primogeniture and hereditary succession, the Mughals also asserted the right of recognising a younger son or even a kinsman as successor. On the death of Raja Man Singh of Amber, Jahangir overruled the claims of Maha Singh, the son of Man Singh's eldest son; Bhao Singh, a younger son of Man Singh, was given the *tīka* instead.

The Mughals levied no tribute, but they expected substantial *peshkash* or "offering" at the time of a ruler's accession or at the time of grant of a *jāgīr*.[4] In assigning *jāgīrs*, the Mughal administration took considerable care in fixing the *jama*' (notional revenue-figure) and tried to obtain full particulars for one or more years regarding the area under cultivation, crops raised, etc.[5] The *peshkash* appears to have been determined with some regard to the difference between the actual collection expected and the *jama*' fixed on the *jāgīr*.

The Mughal emperors succeeded to a great extent in compelling the autonomous chiefs to implement the imperial Mughal regulations, especially in regard to the maintenance of law and order and the freedom of transit. The emperors required the chiefs to take action against rebels, or other offenders who happened to enter their territory. Beyond this they also claimed the right to dispense justice to those who appealed to the Imperial Court against their own chiefs.[6] They even issued

3. *Parwana* (Hindi) *VS* 1765/1708. *Cf. Vir Vinod*, Vol. III, pp. 767-778.
4. Miscellaneous Persian Papers H. 1116 *VS* 1761/1704. The Mahārāja informs the *shāhzāda* that he would pay Rs. 20,000 to him as *peshkash* on receipt of *sanad* for the grant of *nakkārah* (drum) and the *parganas* of Dausa, Chatsu and Mauzabad in *jāgīr*.
5. *Amber Records, Miti, Phalgun Vadi* 10, *VS* 1721/1664. Sabha Chand and Abhai Raj informed *dīwān* Kalyan Das that "a demand has been made from the emperor to send a *taqsīm* for 10 years;" and further that "Gopinath (*qānūngo*) has submitted the *taqsīm* for three years first, and, then, for seven years."
6. *Khutūt* Mahārāja, *VS* 1749/1692; *Parwāna VS* 1768/1711, *Miti, Sawan Sudi* 6. It refers to an *arzdāsht* of *Miti, Sawan, Vadi* 3 same year saying that "a complaint was lodged with Princes Rafiushshan and Jahanshah that the Rāja had been collecting *peshkash* from the people. You may convince the Princes that this is not true and the complaint is baseless."

instructions to the Amber ruler's officials prohibiting collection of forbidden imposts in the ruler's *jāgīrs*.[7] There is considerable evidence to indicate that the Rājas had made repeated efforts in instructing their *pargana* officials through the *dīwān* to avoid complaints going to the Mughal Court so that these might not invite direct interference of the Imperial Government.[8]

Having been enrolled as a *manṣabdār*, the Kachhwaha chief was bound to render military service to the Mughals. The troops supplied by the Rāja (clan contingents as well as mercenaries) contributed appreciably to the military might of the Mughal Empire. Since larger availability of such troops could lead to an increase of *manṣab* and so additional *jāgīrs*, the Kachhwaha chiefs tried to impress their overlord with a show of their military strength. There is an interesting *parwāna* of the Mahārāja (Jai Singh Sawai) to Sah Ram Chand issued at the time when the *Pādshāh* (emperor Bahadurshah) was passing through the adjoining parts of Amber, on his way to Ajmer; the *parwāna* instructs the local officials to collect the troops of Rajawat, Shekhawat, Narooka etc. in order to impress the emperor.[9]

The Mughal Court tended, on the one hand, to treat the Kachhwaha rulers as great nobles of the Empire, and, on the other, insisted on their rendering allegiance personally in acceptance of Imperial paramountcy. Thus it was required that they should remain in attendance at the Court of the

7. *Parwāna* (Persian) 19th *Jama'di-us-Ṣānī*, 2nd R.Y., (6th April, 1721 A.D.), Old Historical Records, hereafter referred to as OHR. From Mughal Court to the *gumāshtas* and *faujdārs* of *pargana* Naraina, Amarsar and Manoharpur, directing them "not to demand the prohibited cesses," such as '*peshkash*,' '*nazrāna*' and '*farmā'ish*' in the *parganas* Naraina, Amarsar and Manoharpur which were in the *jāgīr* of Mahārāja Sawai Jai Singh.
8. *Amber Records*, *Miti Chait Vadi* 3, *VS* 1722/1665 from Kalyan Das and Hathi Ram to *dīwān*, Kalyan Das, regarding the exemption of '*takī*,' for which the *ra'īyat* had gone with a plaint (*faryād*) to the emperor.
   See also *Amber Records*, letter from Kesho Rao to Megh Raj *VS* 1722/1665.
9. *Parwāna* (Hindi) *Miti*, *Mangsir Vadi* 2, *VS* 1764/1707.

Emperor, or should be represented there by one of their close relatives if they themselves held posts elsewhere.

The twin aspects of assertion of 'paramountcy' and recognition of the status of the Kachhwahas were reflected in the Mughal policy of matrimonial alliances. The Amber chiefs were among the few families of the highest nobility in which the princes of the Mughal imperial family continued to contract marriages in the 17th century.[10] Jahangir accorded a special recognition to the close family relations with the Kachhwaha rulers by designating them *Mirzā* Rājas.[11] (The Mughal emperors being descendants of Timur were '*Mirzās*'.).

The support of the Mughal Court greatly enhanced the power of the Kachhwaha rulers. The prestige acquired by them as a result of imperial favour made them stand out prominently among their peers. They, therefore, remained traditionally attached to the Mughal cause; but this did not prevent them from using privileges secured from the Mughals to expand their territories and develop into an independent power upon the decline of the Empire in the 18th century.

II

Expansion of Kachhwaha Territories in Eastern Rajasthan

When Bharamal made his submission to Akbar, his own domain or *waṭan* apparently consisted of the single (though large) *pargana* of Amber. When the *A'in-i Akbari* was compiled (1595), the Amber *pargana* was assigned a *jama'* of 1,22,56,297 *dāms*, and it contained a measured area of 11,35,095 *bīghas*.[12] Both figures are larger than those of the *pargana* of Ajmer, the

10. The marriage alliances with Rajput princely houses in Akbar's reign were, of course, very numerous; but after him the Mughals seem to have adopted a far more selective policy. Cf. Afzal Husain, "Marriages Among Mughal Nobles as an Index of Status and Aristocratic Integration", *PIHC*, Muzaffarpur, 1972, pp. 304-313.
11. The *Tuzuk-i Jahāngīrī* Vol. I, tr. Alexander Rogers; ed. H. Beveridge, p. 266.
12. *Ā'īn-i Akbarī*, ed. Blochmann, Vol. I, p. 508.

capital of the *ṣūba* and headquarters of the *sarkār* within which Amber was placed. However, the *jama*ʻ of this *pargana* could account for only a part of the emoluments to which Bharamal and his successors, Bhagwant Das and Man Singh, became entitled owing to their high *manṣabs* under Akbar. Man Singh rose to hold the *manṣab* of 7000 *zāt* and 7000 *sawār*, and was given large *jāgīrs* in the Panjab and then in Bengal where he built the great city of Rajmahal (Akbarnagar) in 1596.

Man Singh does not seem to have shown any particular inclination towards seeking *jāgīrs* in close proximity to Amber. This may possibly have been due to imperial policy, which might have discouraged additional assignments to Rajput chiefs within Rajasthan, or because the *jāgīrs* outside Rajasthan might have yielded larger revenues in comparison to their *jama*ʻ or, again because of Man Singh's ambitions to enjoy eminence on the imperial stage.

The prestige of the Kachhwahas reached their climax under the celebrated Mirza Raja Jai Singh (1621-1667). Besides the *tankhwāh jāgīrs* that he obtained outside Rajasthan, he was assigned some *parganas* near Amber. In 1627, he was assigned *pargana* Chatsu in *jāgīr*.[13] It was contiguous to his original *waṭan* (Chatsu stood at a distance of "12 kos" from Amber), and is said to have contained his retainers or clansmen (*ulūs*).[14] In 1650, Jai Singh held *jāgīrs* worth 820 lakh *dāms* comprising *parganas* of Amber, Chatsu, Phagi, Mauzabad, Jhak, Behrana, Pachwara,[15] Khori, Deoli Sanchari, Bharkol, Jalalpur, Umran Sakras, Bawal and Jat Kalyawa.[16] Aurangzeb conferred upon him the *manṣab* of 7,000/7,000 in 1661 and 7,000/7,000 (2-3h) in 1665, the maximum rank permitted to any noble. It has been stated in the official history that he could be honoured

13. *Nishān* 1627, with the seal of Nūr Jahan, to Maharaja Jai Singh regarding the grant of a *manṣab* of 4,000 and *jāgīr* of Chatsu. *OHR*, Reg. no. 28.
14. This distance is stated in *arẓdāṣht Miti Mah Vadi* 4 *VS* 1750/1963. See Map 1, facing chapter I.
15. Pachwara included *parganas* Geejgarḥ, Suner, Lalsot, Salaod, Toda Bhim, Liwali and Behroar. *Arhsatta pargana* Lalsot.
16. *Pargana* Jat Kalyawa appears to be Chal Kalyana. C.U. Wills' *The Report On The Land Tenures and Special Powers of Certain Thikanadars of the Jaipur States*, 1933, p. 20; RSA Bikaner.

further only by *in'ām* assignments, which too Jai Singh received.[17] But the *jāgīrs* were not necessarily deemed as his *waṭan*; indeed, they were recognised to be assigned for short durations only.[18] In any case the assignment was absolutely on a personal basis against the high *manṣabs* given to Jai Singh; and the Mirza *Rāja* seems to have made no deliberate attempt to absorb them into his *waṭan jāgīr*. Some of the *parganas* such as Lawan, Paparda and Sunery were held by him in mortgage,[19] i.e. not as *jāgīrs* directly given by the emperor, but on lease from the Imperial assignees of those *jāgīrs*, who had been given loans by the Mirza Raja.

Jai Singh's son, Ram Singh, after becoming Rāja, held a lower rank, 5,000/5,000, and had a restricted area in *jāgīr*, comprising *parganas* Amber, Bahatri (Baswa), Niwai, Hindaun, Phagi and Nihar.[20]

17. *'Alamgīrnāma*, ed. K. Hussain and A. Hai pp. 618 and 917. Bib. Ind. Calcutta, 1865-73.
18. *Sanad* (Persian) 1649, Reg. no. 12; Also see *Fehrist deh-ba-deh Kartik*, *Sudi* 5, *VS* 1720/1663.
19. *Muchalka*, (Bond, in Hindi), 1663, Reg. no. 10.
20. *Arhsattas parganas* Bahatri, Niwai, Hindaun, and *Phagi VS*. 1726/1699, 1738/1681, 1729-32, 35/1672-75, 78 and 1729/1672 respectively. Also see *sanad* (Persian) dated 1656 register No. 25; *Ma'āṣir-ul Umarā'*, II, pp. 301-303.
    Pargana Dausa was held in *jāgīr* by Nawab Bahadur Khan who gave it on *ijāra*. Later on, it was held by the Mahārāja Mirza Raja Jai Singh. But again it was resumed into Royal *Khāliṣa sharīfa* and assigned to Kambakhsh; *Amber Records*, *VS*. 1743/1686.
    The twenty *parganas* which Ram Singh held in lieu of his salary as a *manṣabdār* and which had been attached by Aurangzeb after Shivaji's flight were restored to him upon his accession to the *gaddī* of Amber. M.L. Sharma, *History of the Jaipur State*, pp. 123-124. Sharma does not mention any source for it. He enumerates the *parganas* as follows:
    a. In the Agra District—Audim, Songar (Sonkhar), Songri and Bari.
    b. In the Rewari District—Bawal.
    c. In the Narnol District—Baroda (Rana) and Chal Kaljana (Chal Kalyana).
    d. In the Tejara District—Diwani.
    e. In the Ranthambhore District—Malarna, Newai and Barwara.
    f. In the Alwar District—Jalalpur, Barohar, Patan, Bharkol, Kholari, Malakheda, Mandavar, Kotputli, Thana Ghazi and

In 1688, Raja Bishan Singh was granted *parganas* Amber, Baswa, Phagi, Nihai and Sobner (Jobner) in *jāgīr* with the expectation that he would be able to suppress the Jāts.[21] The *pargana* of Au, home (*waṭan*) of the Jāts, was granted to him in *in'ām*.[22] Later on *pargana* Baswa and Phagi were resumed to the Imperial *khāliṣa* and were assigned to a prince of the Imperial family. After some time a *jāgīr* worth 60,00,000 *dāms* in *pargana* Baswa was given to the crown prince (*kunwar*) of Amber, while the remaining portion worth 20,00,000 *dāms* continued in the *khāliṣa*.[23] *Parganas* Malarna and Jhilai (part only, '*shirkat*') were also held by him in *jāgīr* while *parganas* Niwai and Chatsu were recognised to be in his *zamīndārī*.[24]

One significant feature of these *jāgīrs* was that besides being temporary assignments, they did not often cover the whole (*dar-o bast*) of the individual *parganas*; the *parganas* were shared either with other Mughal *jāgīrdārs* or with the emperor's *khāliṣa*. The pattern may be illustrated by the following examples.

I. *Pargana* Lalsot, *sarkār* and *ṣūba* Akbarabad, 1687.

Total *mauẓas* (villages)	146.00
(*i*) Imperial *khāliṣa*	25.25
(*ii*) Imperial *manṣabdārs*	52.25
(*iii*) Maharaja's *khāliṣa*	68.50

II. *Pargana* Malarna,[25] *sarkār* Garh Ranthambhore, *ṣūba* Ajmer, 1690.

Ismailpur.

21. Extracts of the report (Persian) from Kesho Rai (*Durbār's vakīl*) to Mahārāja (Bishan Singh) dated the 19th Rajab, 31st Regnal Year 1688. C.U. Wills', *The Report On Land Tanures*, p. 20, RSA Bikaner.
22. *Ibid.*
23. *Amber Records Asoj Sudi* 14, *VS* 1753/1696. See *Arhsatta pargana* Bahatri (Baswa). The total *jama'* was 80,00,000 *dāms*.
24. *Amber Records, VS* 1756/1699.
25. *Arhsatta pargana* Lalsot, *VS* 1744/1687.

Total villages—150.00 Total *jama'* 1,00,00,000 *dāms*

(*i*) Imperial *khāliṣa* (*Pai'bāqī*[26] and *khāliṣa sharīfa* and to imperial *manṣabdārs*) *mauẓas* 108.00 *jama'* 73,00,000 *dāms*

(*ii*) Mahārāja's *jāgīrs* 42.00 *jama'* 27,00,000 *dāms*.

III. *Pargana* Naraina,[27] *sarkār* and *ṣūba* Ajmer, 1730.

Total villages	—85.00	Total *jama'*		59,04,400 *dāms*
Imperial (*Padshāhī*) *manṣabdārs*	—11.00	,,	,,	6,94,000 *dāms*
Mahārāja's *jāgīr*	—72.00	,,	,,	48,69,400 *dāms*
Attached to *pargana* Mauzabad	—2.00	,,	,,	3,41,000 *dāms*.

26. *Pai'bāqī* was the designation of areas held by the Imperial administration pending re-assignment in *jāgīr*.
Arhsatta pargana Malarna *VS* 1747/1690. See details of *pai'bāqī khāliṣa* and break-up of imperial *manṣabdārs* as follows:

(i) *Paībāqī khāliṣa sharīfa* — 13,24,875 *dāms* for 24 villages

(ii) Imperial *manṣabdārs* — 59,75,125 *dāms* for 84 villages

Break up:

Name of the Imperial *manṣabdār*	*dāms*	Number of villages held in *jāgīr*
1. *Jāgīr* Inayat-ullah Khan Saif Khan deceased	5,39,152	7
2. *Jāgīr* Sayyed Inayatullah, *qiladar* Khandar	4,69,580	8
3. *Jāgīr* Mirza Fateh Ali Beg and Mohd Qasim	10,42,000	18
4. *Jāgīr* Bahadur Ibrahim	9,19,000	15
5. *Jāgīr* Mirza Mohd	2,30,719	5
6. Mirza Abdullah	94,637	4
7. Sayyed Sadruddin	4,00,000	4
8. Mirza Ismail and Mohd. Quasim	8,10,000	15
9. *Jāgīr* Sayyed Fazil	14,70,037	15

27. *Arhsatta pargana* Naraina, *VS* 1787 1730.

It can thus be seen that the assignment of a *pargana* in *jāgīr* to the Amber ruler did not mean that the Imperial administration also transferred all control over it to him. On the other hand by keeping parts thereof within the *khāliṣa* or assigning them in *jāgīrs* to other nobles or officers, the Mughal government kept parallel sources of authority functioning in the area.

Jai Singh Sawai seems to have been the first Amber ruler to have determined successfully upon obtaining *jāgīrs* in the vicinity of Amber; this may have derived from the difficulty of defending or even governing *jāgīrs* outside Rajasthan in view of the deteriorating political conditions within the Mughal Empire. His further endeavour was to secure the assignment of the *parganas* in entirety (*dar-o bast*); but if this was not possible, he strove to take the unassigned portions in farm (*ijāra*), and thus become the practical master of the whole *pargana*.

At the accession of Jai Singh Sawai, the Kachhwaha principality consisted mainly of *pargana* Amber; and that alone seemed to have been recognized as being under the permanent possession of the Kachhwaha ruler whatever other *jāgīrs* in its vicinity might have been assigned to him.[28] Indeed,

28. *Vakīl's Report* (Hindi) *VS* 1750/1693 from Megh Raj to Mahārāja Bishan Singh informing him that when the *dol* of his *jāgīr* was submitted to the emperor, he granted *parganas* Hindaun, Bayana and Toda etc., to the Mahārāja. The *Vakil* had submitted : "Amber is in the Mahārāja's *waṭan*, the said *maḥals* should also be granted to the Mahārāja." The *Vakīl* was trying to obtain *pargana* Bhusawar also in his (Mahārāja) *jāgīr* as instructed.
Tod in *Annals and Antiquities of Rajasthan*, Vol. III, p. 1351, states that

> "At the accession of Jai Singh, the *rāj* of Amber consisted only of three *parganas* or districts of Amber, Deosah and Bussao; the western tracts had been sequestrated, and added to the royal domains attached to Ajmer. The Shekhavati confederation was superior to, and independent of, the parent state, whose boundaries were as follows. The royal *thāna* (garrison) of Chatsoo to the south those of Sambhar to the west, and Hastina to the north-west; while to the east, Deosah and Bussao formed its frontier."

It is doubtful if Dausa and Bahatri (Baswa) were included in the *watan jāgīr* at the time of Jai Singh Sawai's accession. Probably the Rāja's *zamīndārī* rights prevailed in these *parganas*.

tracts so close to Amber as Chatsu and Manoharpur formed parts of the Imperial reserve (*Badshāhī khāliṣa*).[29]

Jai Singh Sawai extended his *waṭan* by securing tracts in *tankhwāh jāgīrs* and in *ijāra* and *in'ām*. There was a continuous increase in the number of villages assigned to him within various *parganas*, so that his share within them grew steadily larger.[30]

29. *Arzdāsht* (Hindi) *VS* 1750/1693, addressed to Mahārāja Bishan Singh informing him that the writer was trying to obtain *pargana* Chatsu for the Mahārāja in *jāgīr*. The letter indicates that Dausa, Baswa and Mauzabad were not assigned to the Rājā at this time. See also Bhimsen: *Nuskha-i Dilkushā, f.* 170a.
30. The following list of six *parganas* giving numbers of villages held within each by Jai Singh Sawai in various years is illustrative of this process. (Prepared on the basis of *arhsattas*).

Amber (Sawai Jaipur)

Year	*Mauzas (villages)*
1677	541
1715	701
1720	709
1724	780
1725	874
1734	914
1736	947
1737	948
1744	1110

Bahatri

Year	*Mauzas (villages)*
1665	262
1696	320
1718	339
1740	422

Dausa

Year	*Mauzas (villages)*
1709	136
1712	135
1713	135
1741	188
1742	190

Chatsu

Year	*Mauzas (villages)*
1664	150
1708	256
1709	256
1711	289
1712	288
1716	318
1724	326
1726	324
1730	336
1738	399
1741	399
1745	399
1749	400

Some of this increase was due to transfer of villages from one *pargana* to another ;[31] but in other cases, e.g. in *pargana* Malarna (whose limits remained constant) the number of villages in Jai Singh's *jāgīrs* increased clearly at the expense of other assignments.

In 1707, when Jai Singh joined Bidar Bakht near Ujjain, it appears that he had already sent his agent with a request for promotion and grant of *jāgīrs*. The same year he was favoured by Azam Shah with the title of Mirzā Rājā, the *manṣab* of 7,000/7,000 and grant of the *jāgīrs* of *parganas* Chatsu, Mauzabad, Dausa together with such *parganas* as had been held by Mirẓa Rājā Jai Singh.[32] The assignments, of course, lapsed with the defeat and death of A'zam Shah in the war of Succession ; but the grant indicates that such assignments were thought to be a sufficient inducement for Jai Singh to join A'zam Shah. A *parwāna* of 1707, before Aurangzeb's death stresses the importance of securing for the Mahārāja certain areas adjoining his *waṭan*.[33]

Lalsot		*Malarna*	
1687	147	1696	151
1718	149	1708	148
1719	148	1717	148
1734	160	1718	148
1739	169	1720	148
1740	170	1722	148
		1730	148
		1743	153

31. *Miscellaneous Persian Papers*, Register No. 126. R.S.A. Bikaner.
32. *Miscellaneous Persian Papers* (concerning Mahārāja). Also see Satish Chandra's *Party and Politics at the Mughal Court 1707-1740* p. 20.
33. Translation of the Raja's *parwāna* to his *vakīl* at the Mughal Court, dated the 9th Ramzan, 51 regnal year, 1707 quoted from C.U. Wills Report, *Appendix* 'E' : "You wrote to me to obtain the *pai'bāqī* (*Khāliṣa* land) of *ṣūba Dār-ul Khair* (Ajmer) on condition of posting a force there. The *pai'bāqi* of the said *ṣūba* is of no use to us, because it is scattered and far away. It is not a business proposition. The *maḥals* of Chatsu, Mauzabad, Dausa and Niwai are situated close

Subsequently, Jai Singh Sawai succeeded in obtaining transfer of such *parganas* from the Mughal officials to the Mahārāja's *jāgīr*.[34] This was a long-drawn process, because it came into conflict with what was, perhaps, deliberate Imperial policy to maintain control over the area. In addition to the assignments to Mughal *manṣabdārs* in territories adjoining Amber, Imperial officials (*faujdārs*, etc.), who were either Muslims or belonged to clans other than Kachhwahas, were also appointed there.[35]

One way by which this policy was subverted was by Jai Singh Sawai's apparent encouragement to his clansmen to harass the "Turks" (the Mughal Muslim *faujdārs* and *manṣabdārs*), thereby making the latter lose interest in retaining these

to our *waṭan* and *zamīndārī mahals*. The Rajputs, who are to be posted, are already residing there (in the said *mahals*). In these circumstances it is essential to obtain these *mahals*, so that *zamīndārī mahals* may be well administered and also the forces may be kept ready for the service of the emperor as desired. Bearing all this carefully in mind you should act according to the instructions given above. If the *pai'bāqī* is to be given to any agents, you should not be in a hurry to accept . . . you have said that you would obtain the *mahals* of Toda Bhim, Toda Raja Rai Singh etc. near the *waṭan*. It is hoped you will give effect to this proposal ." . . . Also see *Parwāna* (Persian) 1719 *Miscellaneous Persian Papers*. '*Arzdāṣht* (Hindi) from Shah Ram Chand to Mahārāja dated *Miti, Mangsir, Vadi* 2 *VS* 1765/1709, RSA, Bikaner.

34. *Parwāna* (Persian) Register No. 157, 1722, RSA Bikaner. *Parwāna* (Persian) Register No. 195, 1723, RSA Bikaner. *Parwāna* (Persian) dated 15th December 1719. Mughal Court to Samandar Khan ". . . Hastera etc. *pargana* Amarsar, *ṣūba* Ajmer are near Amber ; they must be connected with Amber."

35. Some examples derived from the *Amber records* and *Khuṭūṭ Mahārāja*, may be cited:

A — Pirag Das Rathor—*faujdār* of Bahatri *VS* 1750/1693.

B — Sukr-ullah Khan—*faujdār* of *pargana* Nahar Khoh *VS* 1761/1704.

C — Saiyad Husain Khan—*faujdār* of Amber, *VS* 1765/1708 ; *parwāna Qarār miti Bhadva Sudi* 12 *VS* 1765/1708.

D — Fakhruddin, appointed as *faujdār* of *pargana* Sambhar *VS* 1767/1710.

E — Wahid Khan—*faujdār* of *pargana* Fatehpur *VS* 1769/1712.

F — Nawab Imtiaz Khan—*faujdār* of *pargana* Bairath Singhana *VS* 1775/1718.

jāgīrs.[36] The Mahārāja, after some time, in fact, secured the *faujdārī* and *jāgīrdārī* rights in the territories so disturbed.[37] Nor did the Mahārāja's own officers make things easy for *jāgīrdārs* who shared any *parganas* with him. As early as 1696 such a *jāgīrdār* complained that he was not allowed to collect revenue within his own assignment of *pargana* Chatsu.[38]

It was through leasing of the *parganas* that the expansion of the Jaipur state seems to have mainly taken place.[39] As the Mughal central government began progressively to lose its vigour, the leased areas were practically taken into parts of the Jaipur State. At the same time, however, other circumstances, such as Maratha inroads, or internal disorganisation within the Jaipur territories, also occasioned alterations in the boundaries of the state.

36. *Yāddāsht VS* 1769/1712, *Chiṭṭhī* 1710 from Keshu Singh to *Dīwān* Bhikari Das and Fateh Chand about ousting the *Turks* from the Mahārāja's territories.
37. *Khuṭūṭ-i Mahārāja VS* 1769/1712 No. 510. Draft of *parwāna* addressed to Pancholi Jagjiwan Das acknowledging receipt of his (Pancholi's) '*arzdāsht* regarding the grant of the *nā'ib-subedāri* of Ajmer and the *faujdārī* of Sambhar and Didwana to Nusrat Yar Khan. It directed him to detain Nusrat Yar Khan at the Royal Court till Shyam Singh reached there and met Nawab Hasan Ali Khan to seek the same position and rights for the Mahāraja; and he ultimately succeeded in this.
Also see *Parwāna* (Persian) dated 10th *Shawwāl*, 6th September 1717, 5th regnal year (Farrukh Siyar); *Parwāna* (Persian) from Mughal Court to *chaudhurī* and *qānūngo* dated 14th *Sh'abān*, 3rd Regnal Year (Mohd. Shah), 19th May, 1722 regarding the transfer of *faujdārī* and *jāgīr* of *parganas* Hindaun and Toda from Sa'adat Khan to Mahārāja Jai Singh Sawai.
See *siyāha* (*muḥarrir*) *VS* 1778/1721 for taking the *ijāra* of *faujdārī* of Garh Ranthambhore by Kushal Singh Rājāwāt and Jag Ram Vidyadhar from Muzaffar Ali Rhan in favour of the Mahārāja.
38. '*Arẓdāsht* (Persian) of Bhawani Das (*vakīl-i Durbār*) to Mahārāja' 1696: "that Muhammad Taqi Khan lodged a complaint through *mutṣaddīs* in the Mughal court that he was assigned a *jāgīr* in *pargana* Chatsu worth 1,20,000 *dāms* only; but Hari Ram Sahi and Roop Chand the *gumāshtas* of Mahārāja themselves collect the whole *maḥṣūli* of the said *pargana* and do not allow the revenue-collection ('*āmilāna*) rights to the complainant . . . ." Register 1419, RSA Bikaner.
39. S.P. Gupta, "Expansion of the territories of Amber under Kachhwaha Rulers", *PIHC*, 1965, Allahabad, pp. 177-183.

The crisis in the *jāgīrdārī* system in the later Mughal empire was accompanied by the widespread practice of the Mughal nobles or *manṣabdars* of giving their *jāgīrs* in *ijāra*. Of this tendency the Jaipur house became a considerable beneficiary. In a remarkable letter[40] of 1712 Pancholi Jagjiwan Das, (the Mahārāja's *vakīl* at the Mughal court) informs Jai Singh Sawai that

> "A'itiqād Khan, grandson of Asaf-ud-Daulah has been appointed *faujdār* of Narnaul; and I tried to obtain *ijāras* from him . . . . In compliance with the *darbār's* order I have secured *ijāras* of many *jāgīrs* held by the *Umarā'* (nobles). *Paṭṭas* have been prepared and arrangements for furnishing the surety of some reliable *sahukār* (banker) may be made, so that I may obtain the *paṭṭas* and send them to the *darbār* . . . . The *paṭṭas* of Shujā'at Khan's *jāgīr* consisting of *pargana* of Amarsar, Mauzabad and Bhairana and Nagina under the *sarkārs* of Tijara, have been obtained with effect from Kharif crop . . . . The *paṭṭa* of *jāgīr* of Hasan Khan Ghigada Khan worth 70,00,000 *dāms* in *pargana* Lalsot is ready . . . . The *paṭṭa* of *pargana* Muhammadabad alias Ghazika Thana, held in *jagīr* by Ikhlas Khan is ready. The *paṭṭa* of Mauzpur held in *jāgīr* by Rai Ghasi Ram and Hidayat Kesh Khan *wāqia nigār kul* is ready. The *patta* of Benetta held in *jāgīr* by Jamal Mohd., etc., *manṣabdārs*, and valued at 4,00,000 *dams* is ready . . . . The *jāgīrs* of *Bādshāhzāda* Aizuddin, of Khan Jahān Bahādur, of Azam Khan Bahādur Khān-i Daurān Bahādur, of Khan Jahān Bahādur's son, and of Khan Dauran Bahādur's son are in Mewat. A'ẓam Khān's *mutasaddīs* have refused to give an *ijāra*, but I have won over the *mutasaddīs* of Khān Jahān Bahādur. Khān Jahān Bahādur wishes to establish friendly relations with the *darbār*. The *mutasaddīs* say that the *paṭṭas* for the *ijāra* of the *jāgīrs* of Khān Jahān Bahādur as well as those of his kinsmen will be given as soon as the sureties are produced. The *jāgīrs* are situated close to the *waṭan* and extend from

40. *Vakīl's Report* (Hindi) *VS* 1769/1712. From CU Wills' report, '*Land Tenures*, Appendix 'E', Part II. The text here quoted is in C.U. Wills' MS translation ; the original document remains untraced.

Akbarabad district to Mewat. The acquisition of *ijāra* for these *jāgīrs* will prevent a lot of quarrelling. The holders of other (smaller) *jāgīrs* near the *waṭan* will readily agree to give them to us on *ijāra*. Once these (bigger) *jāgīrs* have been taken on *ijāra* by the *darbār*, I have arranged with the *mutasaddīs* of *Bādshāhzāda* 'Azizuddīn to secure the *ijāra* of *jāgīr* in Perozpur, *pargana* Jhar. I shall report the results later. The *Bādshāhī* courtiers' present policy is to accept whatever price they can get in case the *rājas* do not come to terms with the Emperor and so they are giving out such *ijāras* indiscriminately . . . . Now is the time to take as many *ijāras* as one can. Such *ijāras* will remain in force for three years. If the *darbār* misses this chance of getting the *ijāras* for the *paraganas* close to his *waṭan*, there will be no likelihood of getting these *ijāras* later on. When the Emperor has made a settlement (with the *rāja*), it is doubtful if the *jāgīrdārs* will give out any *ijāra* then."

In the beginning, it appears that the tracts given away in *ijāra* (given for a short period) were a source of profit. But little information is available about *ijāra istimrārī* (*ijāra* on permanent basis)[41] in the early part of the 18th century. It is certain, however, that so soon as the central authority weakened, Jai Singh Sawai acquired some of the territory in *ijāra* for long terms or permanently,[42] and established his full authority there. So long as he served the Mughal Empire as *ṣubedār* or governor, he was interested in retaining the old system of administration (by getting assignments); but when he realised that the Empire was no longer the powerful apparatus that it has been in the past, he began to strengthen his own position around Amber, and through the use of *ijāras* began deliberately to extend his territories. Being himself a high official, he easily secured farming rights in any area which he thought he would be able to control. We have extensive evidence of how a large

41. S.P. Gupta, "Ijara System in Eastern Rajasthan (c. 1650-1750)," *Medieval India—A Miscellany*, Vol. II, pp. 263-274.

42. *Ibid.*

number of *parganas* were taken in *ijara*.[43] He managed these *parganas* by himself leasing them out to sub-lessees or subordinate *ijāradārs*.[44] In the *lakhtang* documents the terms and condition of such contracts have been specifically mentioned.[45] Jai Singh Sawai continued to obtain *parganas* on *ijāra* from the Mughal authorities; and his successors did the same until the Marathas imposed their fiat upon them in 1750.

Jai Singh Sawai had undoubtedly worked out some arrangements with the Marathas whom he had encountered as governor of Malwa (May 23, 1715). Maratha incursions into upper India occurred almost every year from 1732 to 1750, yet his territories remained undisturbed. On *Baisakh Vadi* 8, A.D. 1737 Baji Rao Peshwa wrote to his brother Chimaji: "Sawaiji has also sent us friendly letters requesting us to leave his territory undisturbed . . . we do not disturb his territory as we expect to get supplies of grain and fodder from Sawaiji on our way."[46]

Jaipur thus acquired its position as a major state of Rajputana in the first half of the 18th century as the cumulative result of the position enjoyed by its rulers in the service of the Mughal emperors. The expansion of its territorial jurisdiction was not primarily due to any military exploit but was mainly attributed to the transfer of various *parganas* in course of time by the Mughal authorities to the Jaipur rulers. In the beginning, these transfers were not intended to be permanent and were made in the form of either the ordinary *jāgīr* (assignments) or *ijāra* (leases); but as the Mughal empire weakened, the Jaipur rulers absorbed these territories into their hereditary state. A list of the areas involved is given in the following Table.

After the death of Jai Singh Sawai, Mahārāja Ishwari Singh (1744-1750) held substantial areas in his control. In

43. '*Arẓdāsht* (Hindi) *VS* 1768/1711, Register No. 98. '*Arẓdāsht* (Hindi) *VS* 1769/1712, Register No. 154. CU Wills' Appendix 'E' and 'G'. Also see my article "Expansion of the Territories" *op. cit.* See the table of *parganas* and areas taken in *ijāra* in this Chapter, *infra*.
44. See "Ijara System," *op. cit.*, p. 264. *Muchalka* (Hindi) *VS* 1767/1710. *Bond* (Hindi) Register No. 272.
45. *Lakhtang VS* 1778/1721 and 1781/1724.
46. Irvine, *Later Mughals*, II, p. 248; *Khutūt-i Mahārāja, VS* 1794/1737.

Table[47] showing the Parganas and other areas taken by Jai Singh Sawai in Jāgīr and Ijāra Dāms indicate the Jama' of Parganas in Jāgīr and Rs. indicates the amount of the lease for Parganas in Ijāra)

Year	*Parganas and Other Areas*	*Dāms & Rupees*	*Previous Assignee/Present Assignee from whom lease taken*
1	2	3	4
1699	Malarna	D. 15,40,000	*Pāibāqī Khāliṣa Sharīfa*[48]
1699	Malarna	D. 56,00,000	*Khālīṣa Sharīfa*
1700	Chatsu	Rs. 27,501	*Khāliṣa Bādshāhī*
1707	Lalsot	D. 35,00,000	Chaghtai Khan
1709	Deoti Sanchari	D. 93,77,486	Mughal Court (*Khāliṣa*)
1709	Khohri	D. 197,09,609	Muhammad Khan
1710	Salaod Gurha	—	—
1711	Toda Bhim	Rs. 47,501	Mir Abdullah etc. *Manṣabdārs*
1711	Toda Bhim	Rs. 2,000	Syed Khan
1711	Antela Bhabhra	—	—
1711	Bahroar	—	—
1711	Ghazi Ka Thana	—	Ikhlas Khan
1711	Lalgarh & Tunga	D. 14,90,000	—
1711	Malpura	*Jāgīr*	Rāja Raj Singh Rathor
1712	Bhinai	D. 1,70,000	Bhikan Khan
1712	Khohri	*Ijāra*	Azam Khan

47. This list has been prepared mainly with the help of documents like *sanad*, *yaddāsht*, *parwāna*, *'arzdāsht*, *lakhtang*, miscellaneous Persian papers etc.; concerning the Mahārāja, RSA Bikaner. Cf CU Wills' Report on *Land Tenures*, *op. cit*.

48. *Khāliṣa-i Sharīfa* and *Khāliṣa-i Bādshāhī* are presumably the same.

1712	Khohri	*Ijāra*	Shahzada Rafi-ushan and Futuh-ullah Khan (Lutfullah Khan in the documents)
1712	Mauzpur	*Jāgīr*	Hidayat Khan
1712	Rewari Patodi etc.	—	Khan Jahan Bahadur
1713	Perozpur	*Jāgīr*	Shahzada Aizazuddin
1713	Hindaun	D. 13,60,000	—
1716	Tappa Jhak & Dudhu	Rs. 26,629	Ghulam Ali and Ghulam Zafar
1717	Malpura	D. 1,33,00,000	Aqil-ud-Din
1717	Malpura	D. 46,69,127	Raja Raj Singh
1718	Bhangarh	D. 2,65,100	Wali Mohammad
1719	Amarsar	D. 3,31,066	—
1721	Kiara (Bhangarh)	D. 19,34,000	—
1722	Hindaun & Toda	*Jāgīr*	Sa'adat Khan
1722	Tonk	D. 1,08,00,000	Nusrat Yar Khan
1723	Khohri	D. 1,00,00,000	—
1724	Naraina	D. 95,000	Rahim-ullah
1724	Malarna	D. 44,27,054	Lal Singh
1724	Amarsar	D. 32,00,000	—

1	2	3	4
1727	Village Gogra & Nithari *pargana* Bayana	Rs. 401	Sayyid Abdul Wajid Khan
1727	Pahari	Rs. 2,012	Koke Khan
1727	Alwar	Rs. 1,760/10/-	Koke Khan
		Rs. 663	Dawardad Khan
1727	Manoharpur	Rs. 7,259/8/-	*Khāliṣa Sharīfa*
		Rs. 4,962/12/-	Abdul Baqi Khan
1727	Jhunjhunu	Rs. 17,001	Islam Khan
1727	Tappa Jhak	Rs. 18,281/10/-	—
1727	Naraina	Rs. 157,74/1/-	Sayyid Mir Khan
1727	Kasli	Rs 1,400	Anwer Beg
		Rs. 2,522	*Gurzbardār* Amanullah Beg
1727	Mungana	Rs. 8,286	Laskar Ali Khan
1727	Manoharpur	Rs. 359	Mir Muhsin
1727	Dandikar	Rs. 1,327	Khoja Ahmad Khan
1727	Dehra	Rs. 601	Mukurdi Khan
1728	Toda Rai Singh	Rs. 93,000	Muhammad Khan and Abdul Islam Khan, etc.
1729	*Subah* Ajmer	Rs. 1,75,005/4/-	Obtained from Nawab Muzaffar Ali Khan
1729	Kotputli	—	Nawab Qamruddin Khan
1729	Kotri & Pidawa etc.	Rs. 5,012	—

1729	Bouli etc. (two villages & hamlets)	Rs.	2,036	Sayyid Sarfaraz Khan
1730	Jaitpur & Jhunjhunu	Rs.	10,381	Islam Khan
		Rs.	17,001	Islam Khan
1730	Narhar, Jhunjhunu, Gaonri, Babai & Udaipur	Rs.	1,41,351	Nawab Muzaffar Ali Khan
1731	Malarna & Baboli	—		Mahawat Khan
1731	Gaonri	Rs.	2,600	Khoja Yaqut Abdullah
1731	Tonk	—		Khan Jumma Khan
1734	Kuwawa & Amarsar	Rs.	4,616	Mohammad Shafi etc.
1734	Khandela & Rewasa	Rs.	25,100	Khwaja Qasim etc.
1734	Kotri, Pirana (Malwa)	Rs.	2,956	Jianas Khan etc.
1734	Khandar etc., including Khandar Khandela, Rewasa & Pahari	Rs.	21,059	Zorawar Singh etc.
1734	Khanwa	Rs.	4,490/6/-	Sayyid Nastiar Khan etc.
1734	Khilohra	Rs.	3,612	Kesri Singh and Kishan Singh
1734	Khairwa	Rs.	1,200	Sayyid Wasil Khan
1734	Gayaji	Rs.	1,112/7/-	Mir Hazzayal etc.
1734	Gudha & Sokhar	Rs.	1,874	Γail Ali
1734	Gaonri	Rs.	3,101	Khoja Yaqut etc.
1734	Gijgarh	Rs.	6,608/12/-	Gobind Ram, son of Anand Ram Bankawat

1	2	3	4
1734	Ghusai and Ajai Garh in Malwa	Rs. 1,124/4/-	Taibullah Khan etc.
1734	Nahar Khoh	Rs. 3,055	Ghulam Mehdi etc.
1734	Nagar, Benetha and Uniara	Rs. 4,009/11/-	Fateh Muhammad etc.
1734	Naugaon	Rs. 5,000	Muhammad Rehmatullah etc.
1734	Chandwaji	Rs. 1,406	Munawwar Ali Khan
1734	Chhan	Rs. 3,178	Jamalud-Din Husain Khan etc.
1734	Jalalpur etc. & Umrain	Rs. 12,704	Gulzar Mohammad etc.
1734	Jobner	Rs. 10,477/3/-	Khoja Ros Beg and Ghulam Muztia etc.
1734	Jhunjhunu (Abdul Rahim Khan's holding only)	Rs. 200	Abdul Rahim Khan
1734	Dholpur	Rs. 450	Sabha Singh
1734	Panch Pahar, Subah Malwa	Rs. 7,959/4/-	Kamil Khan
1734	Pindayan, Bharkol and Salawad	Rs. 5,488	Karimullah Khan etc.
1734	Bayana	Rs. 177	Sayyid Fazil Khan
1734	Bheroj	Rs. 790	Baqi Ali Khan
1734	Baletha, Khilohra, Abhaner & Umrani	Rs. 2,553	Khoja Taj Mohammad etc.

1734	Behrouda	Rs.	767	Mahi Singh
1734	Barwada	Rs.	9,500/8/-	Nawaz Beg etc.
1734	Bharkol, Gudha, Sokhar, Maujpur, Alwar, Bhusawal, Salawada & Pindayan	Rs.	19,897	Kunwar Ali etc.
1734	Bahrana, Mauzabad and Udai	Rs.	15,625	Khadim Mehdi etc.
1734	Bhusawal (excluding Badan Singh Jat's holding)	Rs.	15,482	Khusrau Beg
1734	Bhadlao	Rs.	6,332	Khwaja Ahsanullah Khan
1734	Manoharpur	Rs.	1,611	Darwesh Muhammad
1734	Muhammadabad and Baletha	Rs.	19,768/8/-	Mir Muazzam etc.
1734	Malarna, Chhan, Baletha, Bharkol and Lalsot	Rs.	45,336	Qasim Beg Khan etc.
1734	Mauzabad	Rs.	30,544	Khoja Sultan Khan etc.
1734	Mauzpur, Pahari and Bahrana	Rs.	10,785	Khoja Malabad etc.
1734	Richhona	Rs.	9,034/8/-	Mulla Mahmud etc.
1734	Rewasa	Rs.	7,481	Ghulam-ul Nur etc.
1734	Lalsot & Mauzabad	Rs.	1,225	Mehta Bram etc.
1734	Rewali	Rs.	672	Mir Nasr-ud-Din Khan etc.
1734	Udai, Lewali, Malarna, and Amarsar	Rs.	73,338/8/-	Khiv Karan etc.

1	2	3	4
1734	Wazirpur, Ranchhoha and Abhaner	Rs. 14,381	Qudrat-ullah etc.
1734	Salawad, Jalalpur, Lalsot and Rewali	Rs. 12,074	Qulich Beg
1734	Sokhar (excluding Badan Singh Jat's holding)	Rs. 7,705	Qasim etc.
1734	Suneri	Rs. 6,552	Mir Afzal etc.
1734	Singhana, *Subah* Malwa	Rs. 4,510/4/-	Zinas Khan etc.
1734	Harsana & Mauzpur	Rs. 2,294	Nazir Muhammad etc.
1734	Hasanpur, Khohri, Antela & Bhabru	Rs. 32,492/8/-	Ghulam Ahmad etc.
1734	Hasanpur & Madawar	Rs. 3,369	Khwaja Abdullah etc.
1734	Haveli Akbarabad	Rs. 1,009	Jaoli Beg
1734	Antela & Bhabru	Rs. 885	Mirza Ali Muhammad Beg
1734	Ahor, Subah Malwa	Rs. 847	Sayyid Ali
1734	Ajaigarh, Subah Malwa	Rs. 2,286	Ahmad Quli Khan etc.
1734	Ismailpur	Rs. 447	Kalee Khan etc.
1738	Narainpur	Rs. 30,800	Debi Singh, Hukm Singh Shekhawat etc.
1741	Fatehpur	Rs. 24,500	Nawab Muinuddin Khan, son of Nawab Qamaruddin Khan.

1748, he obtained a declaration from the Mughal Court that all the territories that had so far been in the possession of the Jaipur *durbār* would continue in his control in future, and that the friends and foes of the Mahārāja would be regarded as the friends and foes of the emperor.[49] In the beginning Ishwari Singh enjoyed authority over a large portion of these territories, but the civil war between him and Madho Singh greatly weakened his authority.

The limits of the *parganas* which eventually came under the control of the Amber rulers are shown in Maps 1 and 2. These have been drawn up with the help of *arhsattas*, and the reader is invited to consult Appendix, where notes are given on individual *parganas* which were held in *jāgīr* or *ijāra* by Jai Singh Sawai. The surviving *arhsattas* are not in continuous series; and the gaps are considerable. It is, therefore, difficult to show on their basis the precise boundary of the territories controlled by the rulers of Amber in any particular year. However, from the lists of villages in the various *arhsattas*, limits of individual *parganas* can be defined with some confidence. Our maps 1 and 2 are based on detailed maps marking such of the villages as can be located on large scale modern maps. Map 1 delineates the territories held in *jāgīr* by the rulers of Amber c. 1740, with *pargana*-boundaries established from the *arhsattas*. In map 2, a closer view is offered of the six central *parganas*, where the site of each identified village is marked. This may be taken as an illustration of the extent of our knowledge which can, of course, grow as more documents become available. In most of the *parganas* (particularly those which were either large in size or adjacent to *pargana* Sawai Jaipur) about half of the villages have been plotted. Such a large proportion should be sufficient to determine accurately the boundary of each *pargana*. Where identifications have not been possible, this has been due in the main to changes in nomenclature, rather than the disappearance of the villages.[50]

49. *Muchalka* (Persian) year 1748, Register No. 759, from Mughal Court to Mahārāja Ishwari Singh.
50. Sometimes Muslim names have been converted into non-Muslim names; or a small village has been absorbed in a bigger village, the former losing its name; or in case of the foundation of a new village, a new name has been introduced.

As has been explained earlier, the changes in *jāgīr* assignments were quite frequent due to which a *pargana* or portions thereof might constantly change hands. However, with the increasing concentration of Jai Singh Sawai upon expanding his authority in Eastern Rajasthan one can mark a definite tendency towards an increasing number of villages in every *pargana* passing into his control as years passed. Moreover, Jai Singh Sawai, resorted to the technique of taking part of a *pargana* in *ijāra* in the portion outside his own assignment, thereby establishing his full control over the entire *pargana*. In the beginning, some of the *parganas* held by the rulers of Amber in *jāgīr* retained their separate entities, but gradually, with the process of consolidation of the Jaipur state, they were absorbed into other *parganas*. For example, *parganas* Shahpura, Patan, Kaladera, Khoh, Jaitpur, Lalgarh (Lalnagar) and Tunga (all adjoining Amber) were absorbed in *pargana* Amber, which came to be called *pargana* Sawai Jaipur; the absorbed *parganas* were simply treated as its *ṭappas*. The new *pargana* of Sawai Jaipur was established in 1726-27.[51] Similarly, the *parganas* of Bahatri, Dausa, Lalsot, Khohri, Maujpur, and Geejgarh were enlarged considerably at the expense of neighbouring *parganas*. Sometimes such absorption turned out to be permanent; at other times the *parganas* were temporarily attached and grouped with some contiguous *parganas* for the convenience of revenue administration. A *ṭappa* or a *maḥal* or some villages could be attached (*ta'alluqa*) to some *pargana* one year, whereas the same tract was transferred to some other *pargana* the following year. The *arhsattas* often provide us with information on such changes. They give the name of the *ṭappa* with *pargana* containing the villages, a detailed classification of the villages subdivided into *'aṣli* (original) and *dākhlī* (subsequent settlements), and the names of the places taken from other *parganas* (if any) in a particular year.[52] This sort of transfer of villages from one administrative division to

51. *Muwāzana ṭappa ḥavelī, qaṣba* Sawai Jaipur, RSA, Bikaner.
52. *Arhsatta pargana Udai VS* 1793/1736. Total villages 151, out of which 7 were attached to *pargana* Malarna; *Arhasatta pargana* Toda Bheem *VS* 1799/1742. Total villages 223 *ta'alluq pargana* Paota, villages 93, *plus* 130 *ta'alluq pargana* Toda Bheem.

another was quite frequent.[53] Besides the *arhsattas*, other documents such as the *muwāzana* and *taqsīm* papers help us to determine the area of the villages, their locale and limits and the mode of the establishment of a new village.[54]

53. *Amber Record VS* 1742/1685, *arhsatta pargana* Lalsot *VS* 1770/1713. 5 villages of *qaṣba* Sanganer were transferred to Amber, though they continued to remain within the limits of *qaṣba* Sanganer. All villages were in the vicinity of each other. See, e.g. *Arhsatta pargana* Lewali *VS* 1722/1665; also *muwāzanas* of different *parganas*.
54. *Muwāzana ṭappa* Jaipur, *pargana* Sawai Jaipur:
    (i) A village Sursinghpura was previously situated in *qaṣba* Jaitpur but in 1722, it became a village in *tappa* Sawai Jaipur.
    (ii) Village Bakhtpura, *pargana* Sawai Jaipur, had an area of 884 *bīghas*. An addition of 207 *bīghas* was made to this village; and it was renamed Shri Kishanpura. See *Muwazana* of village Bakhtpur, pargana *Sawai* Jaipur.
    (iii) Village Ajitgarh, *ṭappa* Shahpura, Sawai Jaipur, was founded on *Phalgun vādi* 6 *VS* 1803/1746 after the name of Ajit Singh Shekhawat, combining the areas of two villages, namely, Kansubhpura and Andesari consisting of 2,000 *bīghas*, each with revenue Rs. 3,000/- and Rs. 3,500/- respectively. A *fard* was sent to make necessary changes in the papers of *dīwānī fard*. See *Muwāzana ṭappa* Shahpura.

APPENDIX

I

ṢŪBA ĀGRA (AKBARABAD)

Sarkār ALWAR

Parganas	*Year*	*No. of villages in jāgīr*	*Remarks*
1	2	3	4
Antela Bhabhra	1711	16 vill. in *jāgīr*	(9 vill. in Antela, 7 vill. in Bhabhra). Two adjoining *parganas*, often grouped together (*Arhsatta* Antela Bhabhra and Pragpur *VS* 1768/1711).
Ajabgarh	1726	67 vill. in *J.*	In 1756, incorporated in *pargana* Ghazi Ka Thana. The same year *pargana* Bhangarh was treated as a *tappa of* Ajabgarh for a few months.
Bahroj	1690	–	
Barauda	1723	12 vill. in *J.*	Also known as Baroda Mev.
Baraud	–	–	
Bahror	1711	76 vill. in *J.*	–
Bandawal	1724	18 vill. in *J.*	Bandawal (Banawar), previously in *sarkār* Agra. Three *maḥals* (i) Bandawal (*ṭappa ḥaveli*), Harsana and Toda Thék. Major portion of revenue held by Mir Sharfuddin in *jāgīr*.

Baswa	1696	294 vill. in *J.*	Popularly known as *pargana* Bahatri in 17th and 18th cen. Enlarged in *VS* 1797/1740 to contain (a) *ṭappa ḥavelī* (Bahatri), (b) *ṭappa* Murri, (c), *ṭappa* Kolesar, (d) *ṭappa* Chhiri, (e) *ṭappa* Sanchari, and (f) *ṭappa* Rini.
Bharkol	1724	39 vill. in *J.*	
Bairath	1743	33 vill. in *J.*	
Harsana	1723	5 vill. in *J.*	
Hasanpur	1713	11 vill. in *J.*	
Ismailpur	1666	5 vill. in *J.*	*Parganas* Ismailpur and Lahsna often grouped together. Two adjoining *parganas* contained five *maḥals* viz. : (a) Jalalpur vill. 36, (b) Khilohra vill. : 7, (c) Ismailpur vill. 1, (d) Umran vill. 7, and (e) Navroz (Behroz) vill. 1. In 1719, *pargana* Bharkol was also grouped with it.
Khilora	1690	7 vill. in *J.*	
Kotputli	1778	75 vill. in *J.*	Also known as *pargana* Kot (Arhsatta *pargana* Kotputli *VS* 1721/1664).
Lehsana	1666	13 vill. in *J.*	
Mandawar	1724	19 vill. in *J.*	
Maujpur	1723	35 vill. in *J.*	In 1690 contained six *maḥals* viz. : (a) Maujpur, (b) Hasanpur, (c) Khilohra, (d) Harsana, (e) Toda, and (f) Beroz. In 1723, only 5 *maḥals*: (a) Maujpur, (b) Hasanpur, (c) Baroda Mev, (d) Nahar Khoh, and (e) Toda Thek.
Naharkho	1711	17 vill. in *J.*	
Narnol	1765	175 vill. in *J.*	
Pragpur	1740	23 vill. in *J.*	

1	2	3	4
Pinan	1723	22 vill. in *J.*	Also known as Pindayan Pirayan.
Rewari	1665	13 vill. in *J.*	Contained two *ṭappas* viz. : Sheikhpur (vill. 2), and Gurnava (vill. 11).
Rampura	1734	725 vill. in *J.*	Contained 12 *ṭappas* viz. : (a) Amad (vill. 64), (b) Antri (vill. 29), (c) Chandavas (vill. 46), (d) Datoli (vill. 55), (e) Garot (vill. 134), (f) Jalinar, (g) Kohola (vill. 104), (h) Paltar (vill. 70), (i) Rampura (vill. 112), (j) Sanjil (vill. 62), and (k) Shankudhar (vill. 49). *Arhsatta pargana* Rampura *VS* 1791/1734. The state of Rampura, lying on the NW border of Malwa, was an independent state. It was reduced to the status of *zamīndārī* in 1707 with Sangram Singh as its chief under the Maharana of Mewar. In 1728, Jai Singh's wife Chandra Kunwar who was a Princess of Mewar, bore him a son named Madho Singh. Under an agreement, the Maharana granted Rampura as *jāgīr* to Madho Singh, and Jai Singh's agent began to rule this principality. When Madho Singh succeeded to the throne of Jaipur in 1750, he attached Rampura to his state through a *parwāna*. Later on, Madho Singh promised to cede this *pargana* along with *pargana* Tonk to the Marathas (See copy of *parwāna* and *iqrārnāmah* quoted in *Vir Vinod*, II, pp. 973-974 and 976).
Toda	1690	2 vill. in *J.*	Also known as Toda Thek.

Thana Gazi		24 vill. in *J.*	Also known as Ghazi Ka Thana alias Mehmedabad. See (*Chiṭṭhī* of Bhikari Das (Maharaja's *dīwān*) addressed to the Maharaja (date illegible).
Umran Mugan	1713	15 vill. in *J.*	Umran (vill. 7) and Mugan (vill. 8), adjoined *pargana* Jalalpur, in which it was previously included.
Bari Uda (Udai)			In modern maps Bari Uda. In the *arhsatta*, *pargana* Udai, which contained Udai *Buzurg* (Big Udai=Bari Udai), Udai Khurd (Little Udai) and other villages. *Arhsatta pargana* Udai for different years).
Bayana	1693	191 vill. in *J.*	In the documents *pargana* Bayana and Sikandarabad are grouped together (Arhsatta *pargana* Bayana *VS* 1750/1693).
Bhusawar	—	64 vill in *J.*	In 1742, five *parganas* were grouped together, containing 199 *mauzās* viz. : (a) Bhusawar (vill. 64), (b) Toda Bheem (vill. 93), (c) Kathumar (vill. 16), (d) Bayana (vill. 23), and (e) Sonkhar (vill. 3) (*Arhsatta* Paota Bhusawar *VS* 1799/1742).
Dausa	1696	—	In 1709, Dausa was a *ṭappa* in *pargana* Pachwara. In 1736, it appears as a separate *pargana* containing the following *ṭappas*: (a) Riwali, (b) Lawani or Lawano, (c) Suner, (d) Geejgarh, (e) Chharhera, (f) Bhangarh, and (g) Dausa. In 1739, except for Lawani, it contained the other six *ṭappas* : (a) Dausa, (b) Suner, (c) Chharhera, (d) Lawani, (e) Ghazi Ka Thana alias Mehmedabad, (f) Patni, (g) Paparda, (h) Geejgarh, and (i) Behroar.
Guda and Salod (Salaod)	1710	49 vill. in *J.*	Both the *parganas* were generally grouped together (Guda vill. 9) and (Salod vill. 40).

1	2	3	4
Hindaun	1713	214 vill. in *J*.	
Geejgarh	1708	41 vill. in *J*.	Contained three *maḥals* in 1708 : (a) Khuwawa, (b) Suner and (c) Geejgarh.
Kathumar	1742	16 vill. in *J*.	
Khori	1714	241 vill. in *J*.	Contained four *ṭappas* in 1715 : (a) *ḥavelī* (Khori itself), (b) Malikpur, (c) Sikri, and (d) Berher.
Khawa (Khuwawa)	1708	32 vill. in J.	Incorporated in *ṭappa* Patan, *pargana* Sawai Jaipur in 1757.
Mathura	1735	23 vill. in J.	
Lalsot	1740	86 vill. in *J*.	In 1687, a *ṭappa* of *pargana* Pachwara. In *Kāghzāt-i Mutafarriqā*, f. 92 (b), Add. 6586, the *jama*' of Pachwara is mentioned as 3,00,00,000 crore *dāms*. In 1740, a separate *pargana*, it contained three *ṭappas* viz. : (a) Lalsot, (b) Lewali, and (c) Kakrala. Maharaja Sawai Jai Singh got this *pargana* in *jāgīr* as *dar-o bast* (as a whole) in 1726 (*Arhsatta pargana* Lalsot *VS* 1783/1726).
Lewali	1665	22 vill. in *J*.	In 1665, it was a *ṭappa* of *pargana* Pachwara (*Arhsatta* Lewali 1665. See also *arhsatta parganas* Lewali and Kakrala 1772. *Pargana* Lewali contained 26 villages in 1665, out of which 4 villages were attached to *parganas* Malarna, and Udai. The three villages (with 'Muslim' names) were renamed.
Nadbai (Nanadmaee)	1716	—	—
Nagar	1716	42 vill. in *J*.	In *pargana* Nagar, Thuni (*buzurg* and *khurd*) later became headquarters of a separate *pargana*.
Paota	1742	—	—

Sonkhar	1742	23 vill. in *J.*	Attached to *pargana* Paota.
Suner	1708	—	—
Toda Bhīm (Bhiv)	1742	223 vill. in *J.*	Contained 223 villages in 1742, of which 93 were attached to *pargana* Paota and 130 remained with Toda Bhīm (*Arhsatta pargana* Toda Bheem 1799/1742).
Wazirpur	—	—	—
		Sarkār Mathuraji (Islamabad)	
Khoh	1726	102 vill. in *J.*	
Pahari	1737	—	
		Sarkār Kol	
Akbarabad	1732	166 vill. in *J.*	The revenue of almost all the villages was farmed out to local village headmen.
Kol	—	—	—
		Sarkār Sahar	
Sahar	—	—	—

II

Ṣūba AJMER

Sarkār Ranthambhore

Parganas	*Year*	No. of villages *in jāgīr*	*Remarks*
1	2	3	4
Alanpur	1727	4 vill. in *J.*	
Bhagwantgarh	1763	13 vill. in *J.*	*Sarkār* not recorded, but prob. under *S.* Ranthambhore.
Bhadlao	1727	59 vill. in *J.*	Group of three *parganas*: (a) Bhadlao vill. 50, (b) Alanpur vill. 7, and (c) Khilchipur vill 2.
Bhairana	1693	27 vill. in *J.*	
Bhinai	1742	86 vill. in *J.*	*Sarkār* not recorded. In 1744, contained three *ṭappas*: (a) Bhinai, (b) Masuda, and (c) Deo Gaon.
Boraj	—	not known	*Sarkār* not recorded.
Chatsu	1664	244 vill. in *J.*	
Chhan	1734	28 vill. in *J.*	A very important place since the 13th century (written 'Jhain' in the Persian authorities (see S.P. Gupta, 'Jhain of the Delhi Sultanate,' *Medieval India—A Miscellany*, Vol. III, pp. 209-215).
Jhak	1689	16 vill. in *J.*	A separate *pargana* in 1689. Treated as a *ṭappa* of *pargana* Mozamabad in 1693, by 1739 attached to *pargana* Phagi as a *ṭappa*.
Khandar	1771	105 vill. in *J.*	*Sarkār* not recorded.

Kilchipur	1727	2 vill. in *J.*	
Malarna	1730	148 vill. in *J.*	Held by Mahārāja Bishan Singh Ji in 1699 in *jāgīr* as *dar-o bast* (whole) which was rare at this time (See Chapter I). The total revenue (*muqarara tankhwāh dargāhī dām*) of this *pargana* in 1699 was 71,40,000 for 152 *mauza'*s. The break-up follows: (i) *Jāgīr sarkār* (Mahārāja) 5,60,000 *dāms*. (ii) '*Amal ijāra paibaqi khāliṣa*) *sharīfa* lacs 1,54,000. The *jama'* of *pargana* Malarna in *Kāghazāt-i Mutafarriqa*, (f. 77 (a)) Add. 6586, is given as 1,40,00,000 *dāms* in Aurangzeb's time. Up to 1713, the figure of the *pargana* was stated in *dāms*. But from 1714 onwards, the *muqarrara jama'* is mentioned in terms of rupees. The *pargana* officials such as *chaudhurīs* and *qānūngos*, previously Muslims were now replaced by non-Muslim officials. This strongly suggests the continuing authority of the Mahārāja over this *pargana*.
Malpura	1719	150 vill. in *J.*	Also known as Lodri. See '*Arzdāsht*, 1696.
Masuda	1742	52 vill. in *J.*	*Sarkār* not recorded.
Niwai	1701	38 vill. in *J.*	
Nainwa	1665	51 vill. in *J.*	
Toda Rai Singh	1718	128 vill. in *J.*	
Tonk	1727	129 vill. in *J.*	

1	2	3	4
Untgir	1735	17 vill. in *J.*	*Sarkār* not recorded.
		Sarkār Jaipur	
Sawai Jaipur	1744	1110 vill. in *J*	Modern Jaipur, *qaṣba* Sawai Jaipur was founded in *Pos Vadi VS* 1783/1726 by Karim *Multānī* of Sanganer, etc. (*Asāmī* 44), with financial assistance of Amber state. See *Muwāzna qaṣba* Jaipur. The place occupied by the new *qaṣba* was previously known as Chhap Ki Bawri alias Jai Niwas. It became afterwards the seat of the Mahārāja's Court. In 1744, the Sawai Jaipur contained nine *ṭappas*: (a) Amber, (b) Astera (Hastera), (c) Chandwaiji, (d) Jaitpur, (e) Kakar, (f) Khoh, (g) Ramgarh, and (h) Shahpur.
Amber	1714	701 vill. in *J.*	In 1677, *pargana* Amber contained 4 *ṭappas*: (a) *ṭappa ḥavelī* (Amber), (b) Khoh, (c) Kakar, and (d) Ramgarh. In 1677, Amber was placed in *sarkār* and *ṣūba* Akbarabad. In 1724, it comprised of five *ṭappas*, the new *ṭappa* being Hastera. When Sawai Jaipur was enlarged in *VS* 1783/1726, *qaṣba* Amber was made **a** *ṭappa* of Jaipur. (*Muwazna qaṣba* Amber, Sawai Jaipur). According to *Kāghaẓat-ī Mutafarriqa*, f. 92 (b) Add. 6586, estimated revenue (*jama'*) of Amber was 2 crore *dāms*.
Patan	1744	—	Included in *sarkār* and *ṣūbah* Akbarabad in the time of Ishwari Singh, but in 1760 was included in *sarkār qila* (Ranthambhore). *Arhsatta pargana* Patan *VS* 1817/1760.

Shahpur	—	—	Shahpur and Manoharpur were names of the same *pargana*. Manoharpur emerged as a separate *pargana* in 1736 within *sarkār* Nagaur, *Ṣūba* Ajmer, containing two *qaṣbas*, Shahpur and Manoharpur. Subsequently transferred to *S*. Jaipur.
Kala Dera	1747	19 vill. in *J*.	Included as a *ṭappa* in Sawai Jaipur in 1747.
Lalgarh and Tunga	1812		In 1812, the two were treated as *ṭappas* in *pargana* Sawai Jaipur. Lalgarh was also known as Lalnagar.
Kalakh	1744		Contained two *ṭappas*: (a) *ṭappa ḥavelī* (Kalakh), and (b) Borachi.
Naraina	1733	86 vill. in *J*.	
		Sarkār Nagaur	
Khatoo and Amarsar	1732	98 vill. in *J*.	In 1732 *pargana* Amarsar contained two *ṭappas*: (a) Amarsar, and (b) Manoharpur.
		Sarkār Ajmer	
Mauzabad or Mozamabad	1689	77 vill. in *J*.	*Arhsatta pargana* Mauzabad for different years.
Phagi/Fagui	1739	78 vill. in *J*.	Contained three *ṭappas*: (a) *ḥavelī* (Phagi itself) vill. 35, (b) Shri Govindpur vlll. 28, and (c) Jhak vill. 15.
		III. *Ṣūba* SHAHJAHANABAD *Sarkār* Narnol	
Gaonri	1665		

CHAPTER II

Agricultural Production

Rajasthan has been a predominantly agricultural and pastoral region. Desert conditions belong mainly to Western and Northern Rajasthan, but they intrude into the northern portions of Eastern Rajasthan as well. Rainfall is sparse, and no great river (with the exception of the Chambal on its fringe) passes through the region. Sources of irrigation have therefore always been limited. Wells, tanks and natural drainage provided the common means of irrigation. Besides this, bunds were also constructed on slopes or across the rivers, very few of which are perennial. With these limited means of irrigation, the cultivation naturally depended on rains. Uncertainty of rainfall resulted in marked variations in the harvests obtained. *Chiṭṭhīs* sent by the *dīwān* to *pargana* officials (specially *'āmil* and *amīn*) therefore used to be replete with queries about the rains in their respective *parganas*.

Our records for eastern Rajasthan, especially the *arhsattas*, are unique in providing for pre-modern times, a wealth of statistical data which enable us to determine with some exactitude the pattern of cropping and changes in agricultural conditions over time. The following survey is based chiefly on an examination of the *arhsattas* of the contiguous *parganas* of Amber (Sawai Jaipur), Malarna, Bahatri and Chatsu. From *parganas* Dausa and Lalsot also, we have obtained the rates of revenue demand per *bīgha* (see Table II). The period covered in these documents is 1664-1750, but there are gaps because of breaks in the series of the *arhsatta* records.

The *arhsattas* contain information on, among other things, the following :

Total revenue demand; divided into demand from *kharīf* and *rabi'* harvests; cash demand of areas given in *ijāra*;[1] area under each crop assessed according to the *ẓabṭī* (*ẓabṭ*)[2] system, the rate of payment per *bīgha* and the total revenue realised per crop; total revenue realised in kind according to the *baṭ'āī jinsī* system, and the sale prices of the crops in which it was obtained in kind. Sometimes the total produce (*sarwala man*)[3] is also given along with the produce left with the *ra'iyatī* and the share realised by the *dīwān* (state) in kind.

Although information contained in the *arhsattas* is extremely valuable, its analysis is fraught with serious difficulties. In the first place, the area of the *parganas* varied from time to time. For example, the number of villages in *paragana* Amber, later designated Sawai Jaipur, increases from 541 in 1677 to 611 in 1690; 701 in 1715; 780 in 1723; 874 in 1725; 948 in 1737; and 1110 in 1744. This enormous increase was due to the absorption of many contiguous tracts into the *pargana* (Chapter I). In *pargana* Chatsu the number of villages increased from 256 in 1708 to 399 in 1737. In *pargana* Bahatri the number increases from 262 in 1665 to 339 in 1716. In Malarna, however, the size of the *pargana* remained on the whole fairly steady.

Secondly, apart from the fluctuation in area, detailed information is available only in respect of the villages held in the *khāliṣa* (i.e. *khāliṣa* of the ruler of Amber). Even the *khāliṣa* areas, whose revenue was farmed out for a fixed sum, do not offer these returns. The territory held in *khāliṣa* as well as that given in *ijāra* and in *jāgīrs* increased or decreased from season to season and year to year. However, it may be assumed that the crop pattern in the *khāliṣa* areas represented the crop pattern of the territory as a whole.

Thirdly, figures are given in two different categories, the *jinsī* and the *ẓabṭī*, for which the data given are not comparable

1. For the details of *ijāra* See *arhsatta ijāra qarār pargana* Amber of different years.
2. *Ẓabṭ* signified assessment of land-revenue of cash rates separately fixed on each crop for *bīgha*. It implied previous measurement of the land sown with the crops on which these rates were fixed.
3. *Sarwalo man* includes the share of the state as well as of the *ra'iyatī*. *Arhsattas pargana* Malarna, Naraina etc. 1730. A.D.

(quantities and prices in the case of one, and area in the case of the other). As a general rule the crops which are assessed according to the *baṭāī* or *jinsī* system continued to be assessed according to the same system throughout the period. Those assessed according to the *ẓabṭī* system similarly continued to be so assessed as well. The figures for the *ẓabṭī* system do not give any indication of either the produce per *bīgha* or the price of the crop.[4] On the other hand, in the *baṭā'ī/jinsī* system, occasionally the total produce is given along with the quantity demanded as revenue. The government share of produce was, perhaps, sold off on the spot (*bichoti*) and so only the total yield in cash of this, demanded as revenue, is given alongside the sale price of the portion sold.

4. In some *parganas*, the produce per *bīgha* is given : in *pargana* Banawar *VS* 1770/1713 the total area in *rabī'* (*ẓabtī bīgha*) was 543.12.0. Produce per *bīgha*: mans. 9.30.0. The total of *jinsī* (produce) 5299.18.0. (*Arhsatta pargana* Banawar *VS* 1770/1713).
Also see *takhmīna pargana* Gaonri *Vs* 1722/1665. In this document is given the produce of some cash crop per *bīgha* along with the area of foodgrains and their produce.

Crop	*Area bīghas*	*Production Md. Sr. Chatt.*
mung	246	255.10.0
chola	12	10. 0.0
cotton	69	111.28.0
madhwah	4	5.20.0
bajra	800	1243. 9.0
moth	650	722. 0.0
jowar	200	318.15.0
til	50	60.23.4
wheat	317.2.0	3091.34.0
barley	204.2.0	1984. 4.0

In this document the *kharīf* crops were assessed according to *ẓabtī*. Therefore, in *kharīf* the area under *jowar*, *dhan*, *bājra*, *moth*, *mung*, which were ordinarily *jinsī* crops, is also given here. Also see *chiṭṭhī* VS 1776/1719 of *'āmil* to *dīwān*, *pargana* Toda Bhim, where the yield of wheat per *bīgha* is given as : 7.11.0, barley per *bīgha*: 5.2.0; gram per *bīgha*: 1.5.0, *gojro*—5.10.0. Also see *takhmīna kashtkar pargana* Niwai, *faṣal kharīf VS* 1767/1710. The total area of cultivated land was 36,324 *bīghas*. *Ẓabṭ* area was 2993 while the *jinsī* area was 33,331 *bīghas*. Crop-wise break-up is also furnished.

1. Area under Cultivation

The statistical records of the measured area and the assessment made upon cultivated/cultivable land of a *pargana* or a village can be used to determine the proportion of cultivable land (*layaq-zarā'it* Pers. *lāyaq-i zarā'it*) out of the total measured area (*kul-raqba*) (identical with *ārāzī* of the *Āī'n-i Akbarī*) which also includes area under waste (*nābūd*).[5]

It seems that the area figures in our records are given in *bīgha-i daftarī*. The *bīgha* based on it was Two-Third of that *gaz-i-Ilāhī*.[6] The length of the *dorī* consisted of 75 *hāth* (i.e. *kacchi dorī*)[7] and 95 *hāth* (*pakkī dori*)[8] which would make the square based on the former about 62.3% of square based of the latter. The ratio accords closely with that borne by *bīgha daftarī* to *bīgha Ilāhī*.

It seems that in *parganas* for which information is available approximately 7 to 20% of the total measured land was classed as unfit for cultivation (and therefore unassessable).[9] This comprised the *bastī* (area under habitation), *rāha* (tracts, strips), *magro* (rocky lands), *pahar* (hills, if any), *nāla*, jungles, *nadi* (rivers), wells, *tālābs* (ponds), *bagoti* (gardens) etc. It is interesting to note that even the area which was left unassessed was surveyed under different heads.[10] Naturally, the proportion of

5. *Nābūd*, generally was used under the Mughals for the area affected by crop failure. Irfan Habib, *op. cit.*, p. 5.
   However, *nābūd* in these documents has been used for unassessed area, which includes *basti*, wells, trees, *nullah* etc. It means "uncultivable" and, therefore, does not include the area, granted in *punya*, *udik* or to *varhdar* (watchman), *chowkayat* etc. ; which was cultivable but unassessed. See *taqsīms* of different *parganas*.
6. *Taqsīm parganas* Geejhgarh *VS* 1783/1726; Udai *VS* 1800/1743; *Muwāzna dahsāla pargana* Niwai *VS* 1796/1739. Cf. Irfan Habib, *op. cit.*, Appendix A, pp. 353-63
7. *Dastūr-ul-'Amals* of different *parganas*.
8. *Taqsīm pargana* Geejhgarh, *op. cit.*, *Muwāzna dahsāla pargana* Niwai, *op. cit.*
9. S.P. Gupta, "Agrarian Information in *Taqsīm* Documents from Eastern Rajasthan (1649-1767), *PRHC*, Vol. XI, Jaipur, 1979. Reprinted in Hindi in *Madhya Kalin Bharat*, Part I, 1981, entitled 'Purvi Rajasthan Ke *Bhūmi* Sambandhi *Taqsīm* Lekh (1649-1767)', Ed. Irfan Habib, pp. 130-139.
10. *Taqsīm* of different *parganas* in different years. Also *Raqbo pargano* Mauzabad, *VS* 1698/1641.

uncultivable waste to the total measured area varies considerably from *pargana* to *pargana*. It would seem that the total measured area included gross cultivation plus cultivable waste (including *parat* or fallow) and a part of uncultivable waste that is the same as under the Mughals.[11] The actual percentage of area assessed for revenue was between 50% and 65% of the total surveyed land. The rest of the cultivated/culturable land was either *parat* (fallow) or under *sir*[12] or given either in *udik* (charity) or *in'ām*.

Another striking point which emerges from some documents of the years 1665 and 1666, is about the size of the cultivated area (*jot*)[13] out of the total. The proportion of the cultivated area in the total measured area in the villages of *pargana* Chatsu (1666) is found to range from 54% to 81%. In the entire area of the villages for which the data are recorded, the area under *jot* was 55.5% of the total surveyed area.[14] In *pargana* Malarna (1665) the average cultivated area in 34 villages amounted to 79% of the total. The range of variation among the different villages was also not significant. But in six villages the fallow land either exceeds or equals the cultivated area. On the other hand, in 8 villages, the cultivated area accounted for 100% of the "cultivable" area (probably because of the absence of *udik* and *in'ām* lands).

Interestingly a significant proportion of land in a number of villages was under double-cropping. In 6 out of the 34 villages

11. Irfan Habib, *Agrarian System*, pp. 5-6.
12. By an authoritative definition, *Sir* is a term applied to the lands in a village which are cultivated by the hereditary proprietors or village *zamīndārs* themselves as their own special share, either by their own labourers and at their own cost, or by tenants at will, not being let in lease or farm; these lands were sometimes allowed to be held at a favourable assessment, or were unassessed so as to provide *nānkār*, or subsistence for the proprietors: the term is also sometimes applied to lands cultivated on account of the state, or to those in which the revenue is paid by the cultivators without any 'intermediate agent' (Wilson, *Glossary of Judicial and Revenue Terms*, p. 485).
13. *Kishtikār pargana* Malarna 1665; also see *Yaddāshtī haqīqati hal bail zubani patel patwārī pargana* Chatsu 1666.
14. Total number of villages, 78. The total surveyed area, 4,149,68 *bīghas*; area under the *jot* 3,251,51.

of *pargana* Malarna, the cultivated area exceeds the total area (*parat* plus *jot*); this could be only due to double-cropping. Further classification of cultivable land into *polaj* (land under continuous cultivation) and *banjar* (cultivable waste) is also provided in these documents.

Incidentally, it is of some interest to compare the extent of measurement in our period (1649-1767) with that of the closing years of the 16th century, given in the *Ā'īn-i-Akbarī*. The Table below sets out the measured area of six *parganas* in the two periods (the *Ā'īn's* figures are converted into *bīgha-i-daftarī*).

Measured area (in *bīgha-i-daftarī*)

Pargana	*Ā'īn*	*Taqsīm*	Year of *Taqsīm papers*
Antela Bhabhra	37,434	61,180	(1649-63)
Naraina	1,33,307	3,14,378	(1711-22)
Newai	50,890	3,05,967	(1732-41)
Hindaun	7,24,395	6,99,527	(1732-42)
Udai	4,11,100	3,82,543	(1734-43)
Amarsar	1,74,713	94,776	(1758-67)

It appears from the above table that while in three *parganas* out of six, in Hindaun, Udai and Amarsar, the measured area was larger in *Āī'n's* time : in the remaining *parganas* of Antela Bhabhra, Naraina and Newai, the area had increased in the intervening period. We are restrained in drawing any firm conclusion about the relative extent of measurement due to the changes in *pargana* boundaries. However, the general trend towards increase in the surveyed area, and therefore, presumably, in the area under cultivation is unmistakable.

The Principal Crops and Changes in Cropping

From very detailed and valuable documents (*takhmīna sakh unahlu*) of *pargana* Bahatri of *VS* 1797/1740 and *jama-dol* of *VS* 1800/1743 (see Appendix, Table (a), to this Chapter), it appears that the revenue imposed was set as nearly as possible at one-third of the total produce under the *jinsī* system ; but that in estimating the total produce, uniform yields were fixed for each *bīgha*, e.g. 7 *mans* per *bīgha* (Tables b, c and d). The yield so set was for a harvest only. The produce per *bīgha* of barley declined from 8 *mans* in 1740 to 7 *mans* in 1743. Prices fluctuated from year to year. Whereas the wheat price tended to be *mans* 1.10.0 per rupee in 1740, it had risen to *mans* 1.5.0 in 1743 (Appendix tables b and c to this Chapter).[15] The price of barley rose from 2 *mans* per rupee in 1740 to *mans* 1.25.0 in 1743, perhaps in response to an indifferent harvest of barley in the latter year, indicated by a decline in the officially estimated yield just mentioned. The most valuable point in these documents is that these contain particulars of the total area under cultivation, total produce, state share in kind, the price at the cash value of the state share. If more such documents became available, they would help us to analyse the nature of productivity, the extent of cultivation under various crops and the changes in the incidence of revenue.

But the bulk of the *arhsattas* give us a much more limited range of information. For crops under the *batāī* system they usually give the total produce, the state demand in quantity, the price and the cash value of the state share. In the *ẓabṭī* system, we have the total area under each crop, the rate of revenue demand per *bīgha* and the total revenue demand in cash. Thus owing to the absence of information about the total area in the case of *jinsī* crops and about the total quantity of production in the case of *ẓabṭī* crops, our data for most *parganas* and most years tend to be incomplete.

In view of these limitations of evidence in our study of the relative position of crops, only one common denominator has been used, namely; the revenue realised from each crop (Table I). This method gives us a rough indication only, because there are too many variable factors involved to enable us to judge the crop pattern exclusively by comparing the revenue derived from different crops. Under the *ẓabṭī* assessment different rates per *bīgha* were applicable even in the same villages, and there

(*Text. continued on p. 50*)

15. Cf. S. Nurul Hasan and S.P. Gupta, "Prices of Foodgrains in the Territories of Amber (c. 1650-1750)". *PIHC*, Patiala, Part I, pp. 345-368.

Table I

Table showing the Percentage of Revenue Derived from different Crops

Kharīf

Pargana Jaipur

Year	*Bajra*	*Jowar*	*Moth*	*Urd*	*Mung*	*Chola*	*Sugar-cane*	*Cotton (Vani)*	*Maka*
1666	31	17	27	—	1	—	1	13	—
1677	32	3	35	1	1	—	3	16	1
1688	18	1.8	27	1	3	—	1.3	18	—
1689	18	—	37	—	3	—	—	11	5.4
1690	11	1.4	41	1	3	—	—	24	2.5
1712	31	1.2	32	—	2.5	—	2.7	14	5
1715	15	4.6	18	1	1	—	12	19	12
1716	5.7	25	18	1	1.5	5	—	19	15
1718	30	5	13	—	—	2.5	3	15.6	12.6
1720	25	1.8	15.4	1	1.8	5.6	3	16	16
1723	22	3	20	—	—	6	1.4	20	12
1724	8	2	27	—	2	8	3	15	15
1725	13	1	7	—	—	8	5.6	25	20
1733	3.6	1	10	1	1.4	1	3	46	12.4
1734	4.3	1.7	26	1.9	2.3	9.1	6	19.8	16
1737	6.3	1.7	23	1.9	1.8	11	4	20	18
1744	7.7	8	20	—	1	6	10	9.8	17

Pargana Malarna

Year	*Bajra*	*Jowar*	*Moth*	*Urd*	*Kodon*	*Sugar-cane*	*Cotton*	*Maka*
1690	13.2	26.8	19.6	—	1	23.6	9	—
1699	29	20	6.2	1.5	—	16	7.8	—
1711	40	26	9.8	—	--	10	1.5	—
1712	32	15	14	—	—	—	—	—
1713	35 6	9.3	26.9	5	1	5.9	2	—
1714	41.8	11	6.8	2.7	1.9	15.3	3.9	—

Contd.

Year	Bajra	Jowar	Moth	Urd	Kodon	Sugar-cane	Cotton	Maka
1715	34	15.5	14.2	3.7	1.8	14.9	6.2	—
1716	26	5.2	17.7	7.7	1.7	14	7.4	—
1717	37.3	11.2	14.7	9.2	1	7	7.5	—
1718	42.6	16.9	6.3	2	3.3	9	10	—
1719	24.9	29.3	18	2.2	2.7	8	—	—
1720	32	29.3	7.3	1.6	1.3	10.6	4.2	—
1722	36.8	26.6	8	1.6	—	7	4.2	—
1723	34.3	33.8	8.8	1	1	5.4	5.2	—
1726	21.6	29.6	3.8	2.3	3.2	20	9.7	—
1727	27	30	6.6	3	2.7	10.2	6.9	—
1728	21.8	24.4	11.8	1.9	2.3	4.3	8.8	—
1729	22.8	26	9.3	2.6	2	9	13.7	—
1732	16.5	26	3.3	1	6.3	12.8	22.2	—
1736	15	25.3	5.5	3.9	1.7	20.4	22	—
1737	19.5	30	9	1.5	3.7	11.6	12.3	—
1738	28	31.6	14.2	2.3	1.6	3	2.5	—
1743	30.3	38.2	3	—	—	5.3	5.7	—
1744	21.7	26.5	15	—	2	14.6	2	—
1745	29.2	25.2	10.2	2	—	7	5	—
1748	26.8	27.8	12.6	2.5	—	5.5	4	—
1749	19.4	34.8	10.2	2 4	—	6.4	2.8	—

Pargana Bahatri

Year	Bajra	Jowar	Moth	Urd	Mung	Sugar-cane	Cotton	Maka
1665	14.8	15	13.7	2	1.3	—	17	—
1669	16	21.4	15.3	4.8	2.5	2	27	—
1684	11	15	18.4	8	2	5	30	—
1685	11	13	9.6	4	—	7.8	44	—
1686	15	15	15	4	—	6	27	1
1688	4.7	14	13	5	—	8	28	2
1689	—	17	22	6	—	21	16	—
1691	4	16	20	5	—	9	30	—

Contd.

Year	*Bajra*	*Jowar*	*Moth*	*Urd*	*Mung*	*Sugar cane*	*Cotton*	*Maka*
1696	10	13	2	—	—	9	33	7
1706	24	23	17	—	—	5	19	1
1708	27	25	12	1	—	5	16	3
1710	12	13	32	7	4	3	18	2
1716	5.3	18.7	13.8	8.9	11	12.8	33	2.4
1717	10.4	5.6	3.4	—	—	29.8	29.9	2.8
1718	24.4	18	12.7	2.5	—	2.7	17.2	4.6
1720	12.2	6.7	13	4.1	2	14.4	23	9.9
1721	14.6	15.9	15.7	5.8	1.6	8.1	15.3	6.2
1723	17	17.6	13.8	27	—	5.1	23	6.2
1724	10	10	18.9	4.5	1.2	6	13.1	5.6
1725	8.6	5.9	16.4	10.3	1.2	10.5	25.4	6.8

Pargana Chatsu

Year	*Bajra*	*Jowar*	*Moth*	*Urd*	*Kodon*	*Chola*	*Sugar-cane*	*Cotton*	*Maka*
1664	54	8.4	13	—	1.6	—	—	7	—
1708	55	5	1	1	—	—	—	—	—
1710	40	7.6	26	—	3	—	—	5.5	1.2
1711	51	3.4	16	—	4.3	—	—	10	1.8
1712	56	3.8	15	—	3	—	—	10	2.3
1713	11	12	35	4	—	—	5.2	6.2	4.4
1714	35	8	28	—	9	—	—	9	4.5
1715	41	7	26	1.9	4.5	—	2	7	—
1716	24	—	22	5.2	9	—	—	19	—
1721	35	9	21	2	4.8	1.8	1.8	10	5.5
1723	36	12	10	—	4.2	1.2	—	18	7
1724	15	11	37	1.2	3.7	3.3	1	9	2.3
1726	11.4	7.8	10	2.7	1.7	4	6.4	42	3.4
1730	10	1.8	13	3	10	1.8	6.3	32	6.5
1731	5.7	1.8	—	—	—	—	8.3	56	14
1733	7.8	2.3	19	3.4	9.8	2.7	4.2	24	8.6
1735	3.3	—	28	7.4	6.6	5.2	4.6	19	2.2
1736	14	7.5	6.8	2	5.8	5.5	4.5	31	12
1737	6	4	27	2.5	7	5	3	23.5	9.5
1740	7.7	4.8	19	3.7	9	4	2.4	20	11
1741	5.4	1.6	12.4	3.2	9.4	1.9	6.8	29.5	12.2

Contd.

Year	Bajra	Jowar	Moth	Urd	Kodon	Chola	Sugar-cane	Cotton	Maka
1742	11	10	3.2	—	4.7	1	8.9	40.4	11.3
1744	10.3	13.5	25.6	2.4	7	2.7	3.4	5.7	3.5
1745	14.1	10.3	16.2	3	7 3	2.9	3	11	12.8
1748	19	2.9	13.8	1.4	4.9	2.2	—	21.6	14.7
1749	11.5	4.5	14.4	2.8	5	2.1	1.7	18.8	19.2
1750	4.6	3	7.4	3	—	—	3.3	28.3	25.6

Rabī'

	Pargana Jaipur			*Pargana Malarna*		
Year	Wheat	Barley	Gram	Wheat	Barley	Gram
1665	—	—	—	39	29	21
1677	5.1	49.6	40	—	—	—
1688	15.8	52.9	27	—	—	—
1689	10	57	23.9	—	—	—
1690	7	47.5	33	39.4	29	21
1711	—	—	—	11	14	40
1712	21.7	58	12	—	—	—
1713	—	—	—	13	54.7	18.6
1714	—	—	—	19.9	66	9.7
1715	23.7	57	15	14	48	—
1716	31.4	50.6	17	—	53	23
1717	—	—	—	15	71	9
1718	29.6	63.6	1.2	17	69	11
1719	—	—	—	17	53	16
1720	17.4	72	1	25	38	8
1722	—	—	—	15	62	17
1723	16	77.8	—	17	65.5	10
1724	17	29	9	—	—	—
1725	10.5	50	29.5	—	—	—
1726	—	—	—	9.7	43	27.9
1727	—	—	—	18.6	41	28.9
1728	12.6	61.5	6	13.8	35.6	35.7

Contd.

	Pargana Jaipur			*Pargana Malarna*		
Year	*Wheat*	*Barley*	*Gram*	*Wheat*	*Barley*	*Gram*
1730	—	—	—	10.8	38	35.4
1732	—	—	—	8.5	67	20
1733	9.6	52.3	3.2	—	—	—
1734	15.3	73.4	1	—	—	—
1736	—	—	—	1.6	70.7	—
1737	11.7	82	1.6	17.9	68	7.5
1738	—	—	—	24.8	26	21
1743	—	—	—	37.6	45.8	8.5
1744	12.8	73.8	5.9	25.7	47	17
1745	—	—	—	21.8	36.9	19
1748	—	—	—	36.4	45	12.7
1749	—	—	—	34.5	36.5	25
1750	—	—	—	42.6	38	13.5

	Pargana Bahatri			*Pargana Chatsu*			
Year	*Wheat*	*Barley*	*Gram*	*Wheat*	*Barley*	*Gram*	*Chombli*
1665	14 5	61.8	17	—	—	—	—
1669	7.7	60.8	21.7	—	—	—	—
1686	11.7	61.5	11.7	—	—	—	—
1688	20.7	48	9.5	—	—	—	—
1689	24.4	45	8.6	—	—	—	—
1696	13	78.6	—	—	—	—	—
1697	14.7	72.4	4.3	—	—	—	—
1706	22	60.7	11.6	—	—	—	—
1708	21	63	7.8	—	—	—	—
1710	24.8	39	27.4	10	48	31	2
1711	30	34	28	16	41	—	2
1712	—	—	—	15	60	7	9
1713	—	—	—	30	49	14	—
1714	—	—	—	14	49	24	8.6
1716	32	42	18	13	37	27	10
1717	18	73	—	10	78	—	6.7
1718	17	70	3	—	—	—	—
1720	28	67	—	—	—	—	—

	Pargana Bahatri			*Pargana Chatsu*			
Year	*Wheat*	*Barley*	*Gram*	*Wheat*	*Barley*	*Gram*	*Chombli*
1721	25	58	9	15	57	18	2.8
1723	29	59	5	13	75	—	5.4
1724	28	52	11	17	52	14	4
1725	20	47	19	—	—	—	—
1726	—	—	—	10	57	9	12.5
1730	—	—	—	10	59	18	4.5
1731	—	—	—	8	77	—	7.9
1733	—	—	—	14	60	6	11
1735	—	—	—	15	56	12	3.5
1737	—	—	—	10	70	7	3.4
1738	—	—	—	8	63	12	9.8
1740	—	—	—	12	57	9	5.5
1741	—	—	—	13	60	13	4.7
1742	—	—	—	3.2	79.6	1.4	5
1743	—	—	—	8	74	4	6.4
1744	—	—	—	13	66	5	5
1745	—	—	—	13	63	5	11.9
1748	—	—	—	17	65	2	9.6
1749	—	—	—	11	64	7	10

were changes in the rate of demand from year to year. Apparently, the rate depended on the quality of the produce, the productivity of the soil and the current prices. Nevertheless, an examination of the rate of revenue demand per *bīgha* for *ẓabṭī* crops shows a remarkable continuity (Table II). Since during the period under review there was apparently a general rise in prices (see Chapter III); the *ẓabṭī* system would appear to have been more advantageous to the *ra'īyat* than to the state. It was perhaps for this reason that there is a tendency for an increase in the cultivation of *ẓabṭī* crops.

For the *jinsī* crops too, the revenue figures do not necessarily provide a firm guide to the crop pattern. The fluctuations in price offer, of course, an obvious difficulty. At the same time we know

(*Text continued on p. 59*)

16. *Ibid.*

FIG. 1

PERCENTAGE OF REVENUE (RABI HARVEST) DERIVED FROM WHEAT

JAIPUR

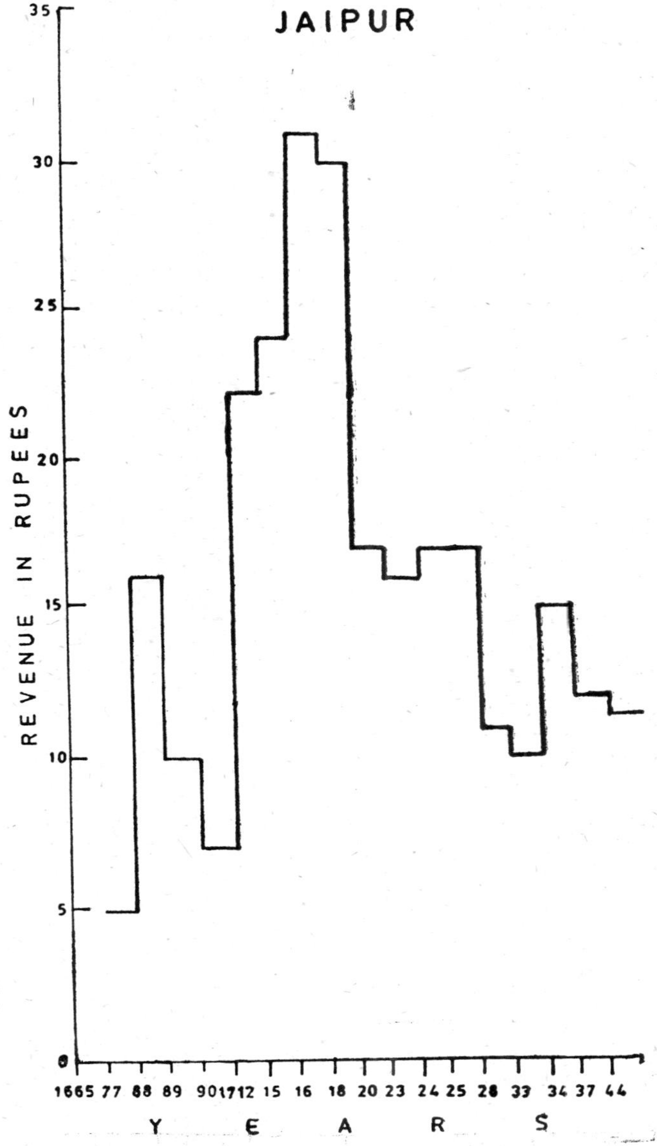

FIG. 2

PERCENTAGE OF REVENUE (RABI HARVEST) DERIVED FROM WHEAT

BAHATRI

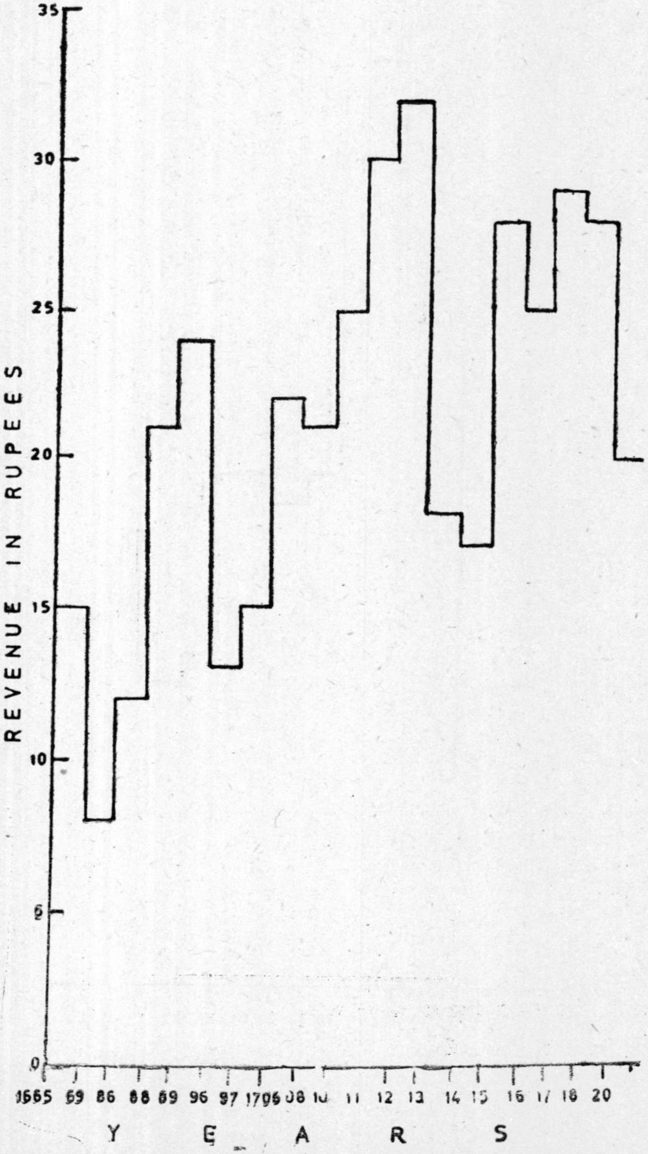

FIG. 3

PERCENTAGE OF REVENUE (RABI HARVEST)
DERIVED FROM WHEAT
CHATSU

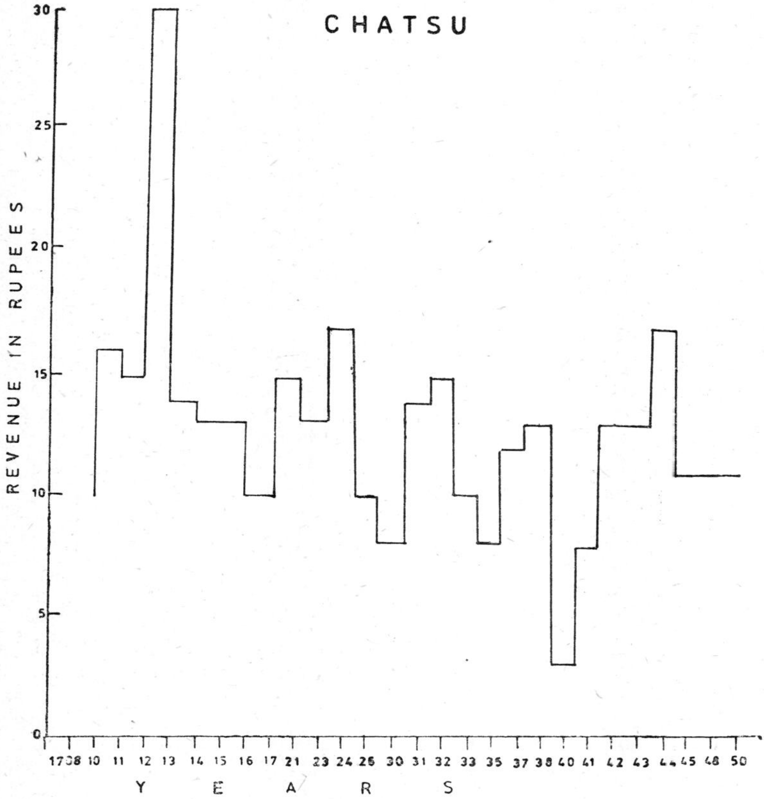

FIG. 4

PERCENTAGE OF REVENUE (RABI HARVEST)
DERIVED FROM WHEAT

MALARNA

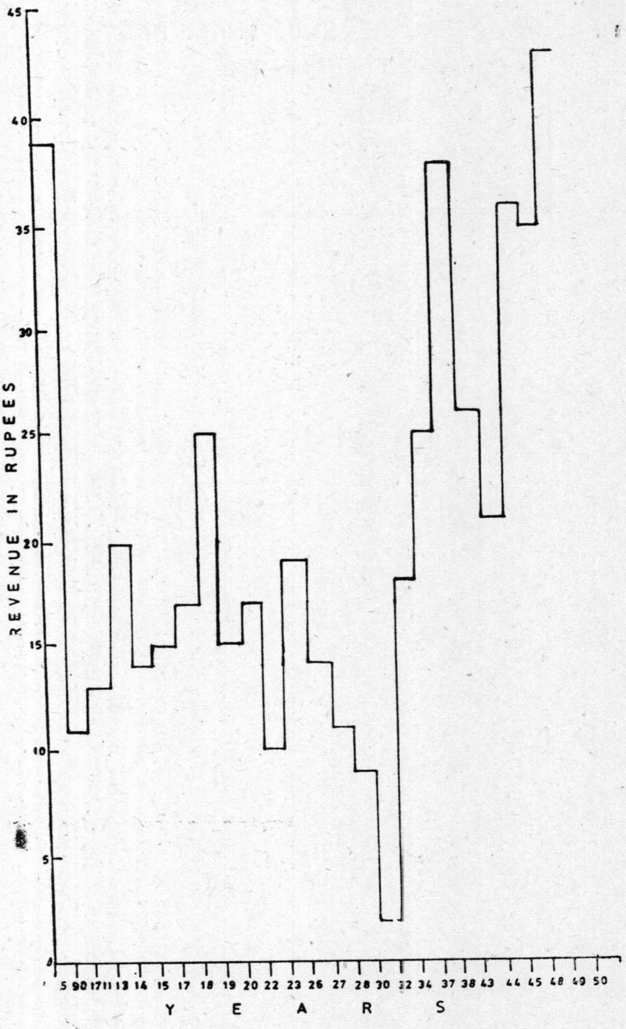

Table II

Rate of Revenue Demand Per Bīgha

(In Rupees)

Pargana Jaipur

Year	*Cotton*	*Sugar-cane*	*Kodon*	*Maka*	*Chola*
1	2	3	4	5	6
1666	1.28	4.73	1.17	1.87	0.94
1677	1.35	4.64	1.17	1.61	0.63
1688	1.28	3.70	1.14	1.57	0.62
1689	1.33	—	—	—	0.60
1690	1.36	3.75	1.16	1.50	0.61
1712	1.13	2.62	1.08	1.28	0.49
1715	1.23	3.31	1.10	1.42	0.48
1716	1.18	—	1.07	1.34	0.49
1718	1.21	3.19	1.12	1.37	0.49
1720	1.20	3.22	1.03	1.28	0.50
1723	1.23	3.30	1.08	1.31	0.51
1724	—	—	1.08	—	0.50
1725	1.27	3.32	1.09	1.36	0.52
1728	1.23	—	1.00	2.00	0.64
1729	1.16	3.91	1.00	1.69	0.60
1733	1.18	3.74	1.11	1.46	0.57
1734	1.29	3.68	1.12	1.45	0.56
1736	1.29	3.69	1.11	1.42	0.57
1737	1.30	3.77	1.11	1.44	0.56
1744	1.36	3.68	1.11	1.41	0.56
Pargana Malarna					
1690	1.24	3.24	0.97	—	1.00
1699	1.09	2.69	1.00	0.87	0.83
1711	1.22	3.15	0.99	1.00	1.00
1713	1.21	3.11	0.96	0.94	0.96
1714	1.21	3.06	0.98	0.98	1.00
1715	1.15	3.01	0.95	1.00	1 00
1716	1.22	2.96	0.98	1.00	0.98
1717	1.19	2.89	1.00	—	
1718	1.18	3.06	0.93	0.88	1.00

Contd.

1	2	3	4	5	6
1719	1.18	2.86	0.88	0.91	0.99
1720	1.19	3.07	0.98	1.00	0.97
1721	1.21	3.15	0.97	1.00	0.98
1722	1.22	3.13	0.97	1.00	1.00
1723	1.20	3.19	0.97	0.97	0.99
1726	1.21	3.18	0.99	1.00	0.99
1727	1.24	3.24	0.99	1.00	0.99
1728	1.22	3.25	0.99	1.00	0.99
1729	1.23	3.22	0.99	1.00	1.00
1730	1.24	3.16	0.99	1.00	1.00
1731	1.23	3.19	1.00	0.97	0.98
1732	1.22	3.21	0.96	0.94	0.95
1736	1.23	3.18	0.99	0.99	1.02
1737	1.23	3.16	0.99	0.98	1.02
1738	1.23	3.17	1.00	0.99	1.00
1743	1.21	3.02	—	0.98	1.00
1744	1.21	3.14	0.99	0.98	0.99
1745	1.15	3.09	0.97	0.98	1.00
1748	1.17	3.09	0.99	0.98	0.98
1749	1.17	3.11	0.96	0.98	0.98
1750	1.09	3.97	0.99	0.99	0.92

Pargana Dausa

1709	1.29	—	1.00	1.17	0.61
1712	1.34	1.53	1.12	1.38	0.68
1713	1.34	1.42	1.09	1.29	0.69
1715	1.26	1.50	1.12	1.21	0.65
1736	1.43	1.86	1.18	1.40	0.70
1737	1.42	1.97	1.16	1.38	0.70
1738	1.48	1.99	1.24	1.38	1.38
1739	1.39	1.78	1.13	1.36	1.36
1741	1.41	1.69	1.20	1.40	1.40
1742	1.35	1.68	1.14	1.32	1.32
1743	1.31	1.63	1.09	1.28	1.28
1744	1.39	1.73	1.14	1.40	1.40
1745	1.36	1.69	1.13	0.62	1.34
1746	1.41	1.77	1.26	0.65	1.39
1747	1.41	1.80	1.15	0.65	1.39
1750	1.39	1.66	1.18	0.64	1.38

Contd.

Pargana Lalsot

1	2	3	4	5	6
1687	1.02	1.95	0.79	1.53	0.57
1713	0.93	1.72	0.69	0.96	0.51
1714	0.91	1.81	0.74	1.09	0.59
1715	0.94	1.87	0.73	0.94	0.61
1717	0.95	1.86	0.87	—	0.60
1718	0.98	1.69	0.79	0.91	0.59
1719	0.92	1.79	0.81	1.03	0.60
1720	0.93	1.71	0.78	0.96	0.60
1721	0.95	1.72	0.79	0.97	0.60
1722	0.94	1.69	0 76	0.97	0.59
1731	0.98	0.88	0.81	0.86	0.60
1733	0.98	1.79	0.76	0.86	—
1734	0.97	2.22	0.76	0.89	0.60
1737	0.98	1.86	1.77	0.97	0.61
1739	0.97	1.85	0.73	1.01	0.61
1740	0.99	1.80	0.74	1.06	0.62
1741	0.99	1.83	0.75	0.95	0.62
1745	0.92	1.79	0.83	0.87	—
1746	0.99	1.83	0.75	0.86	0.63
1747	0.98	1.85	0.76	1.00	0.61
1748	1.08	1.78	0.70	0.89	0.58

Pargana Bahatri

1	2	3	4	5	6
1665	1.41	2.00	—	1.74	—
1669	1.44	2.00	—	1.79	—
1684	1.45	2.00	—	—	0.93
1685	1.40	2.00	1.60	1.60	1.20
1686	1·44	2.00	1.40	1.23	—
1688	1.30	1.80	—	1.25	—
1689	1.39	2.00	1.50	1.74	1.09
1691	1.25	2.00	1.25	1.80	1.08
1696	1.25	1.75	1.25	1.33	1.00
1706	1.30	1.90	1.40	1.30	1.00

Contd.

1	2	3	4	5	6
1708	1.36	1.83	1.33	1.25	1.23
1710	1.34	1.87	1.44	1.25	1.78
1716	1.30	2.00	1.50	1.20	0.76
1717	1.25	1 91	1.46	1.21	1.09
1718	1.37	1.96	1.46	1.19	0.76
1720	1.34	2.00	1.40	1.23	0.74
1721	1.33	1.94	1.39	1.24	0.74
1723	1.35	1.96	1.40	1.24	0.73
1724	1.42	1.95	1.47	1.24	0.74
1725	—	—	—	—	0.74
Pargana Chatsu					
1664	0.82	2.00	0.83	1.09	0.97
1708	1.24	—	1.12	1.36	0.75
1709	1.23	—	0.96	1.41	—
1710	1.12	—	1.03	0.95	0.47
1711	1.11	2.02	1.00	1.07	0.28
1712	1.11	2.30	0.98	1.08	0.59
1713	1.34	1.96	1.41	1.24	0.76
1714	1.10	2.20	0.95	1.05	0.53
1715	1.11	2.45	0.97	1.10	—
1716	1.11	2.71	0.94	1.08	0.50
1717	1.10	2.59	0.90	0.98	0.45
1721	1.11	2.75	0.97	1.10	0.54
1723	1.08	2.74	0.95	1.08	0.49
1724	1.11	2.91	0.98	1.10	0.55
1726	1.12	2.97	1.08	1.10	0.56
1730	1.10	2.91	0.98	1.08	0.54
1731	1.08	2.82	—	1.07	0.51
1733	1.11	2.97	0.98	1.09	0.55
1735	1.11	2.97	0.98	1.05	0.55
1736	1.07	2.92	0.96	1.05	—
1737	1.09	2.93	0.92	1.06	0.52
1740	1.09	2.94	0.95	1.09	0.77
1741	1.10	2.88	0.96	1.09	0.55
1742	1.10	2.91	0.97	1.09	0.55

Contd.

1	2	3	4	5	6
1744	1.10	2.73	0 97	1.07	0.56
1745	1.09	2.86	0.97	1.06	0.55
1748	1.07	2.44	0.94	1.04	0.54
1749	1.08	2.59	0.96	1.06	0.55
1750	1.10	2.91	—	1.07	—

from these documents that the share of the State in respect of the different crops varied even within each *pargana* and quite frequently from year to year. The share varied in accordance with the fertility of soil, from one-fourth (*vat-chautho*) to one-third (*vat tisro*), two-fifths (*vat panch-duvo*), and to one-half (*vat adho*). It would appear that the standard revenue demand in *jinsī* crops as 2/5ths of the produce on *polaj* land and ¼th of the produce on *banjar* land was universally accepted. Generally, the *paṭṭa* or *vat vih* (share in cultivation) for *banjar* land was given on a payment of ¼th of produce to the state; however, cases of 1/3rd share are also recorded.[17] The following table gives the average of the share of the total produce for each of the principal crops taken by the State in the four *parganas* of Amber, (Jaipur), Malarna, Bahatri and Chatsu.

State Share of Total Produce (in percentage)

Kharīf

Pargana	*Year*	*Bajra*	*Jowar*	*Moth*	*Urd*	*Mung*	*Til*	*Chola*
Amber	1716	38.2	45.2	40	40	38	40.5	—
Malarna	1699	42	44.5	41	44	43	44	—
Bhatri	1665	45	46	44	48	43	42	50
Chatsu	1709	35	36	36	38	39	33	—

Rabī'

Pargana	*Year*	*Barley*	*Wheat*	*Gram*
Amber	1716	29.7	30.5	34.6
Malarna	1699	34	34.6	38
Bahatri	1665	33	32	41
Chatsu	1709	32	33	37.7

17. *Chiṭṭhī Miti Baisakh Vadi* 10, *VS* 1801/1744; *VS* 1774/1717. *Lakhtang* (undated) *Miti Phalgun Sudi* 1, place Jaipur. *Chiṭṭhī Miti Bhadva Vadi* 11, *VS* 1806/1749.

A perusal of the above table reveals that the incidence of demand for the principal *jinsī* crops of the *rabi'* harvest was more or less the same as that of the *ẕabṯī* crops while for the *kharīf* harvest the incidence of the revenue demand of the *jinsī* crops tended to be 10 to 15 per cent higher than that of *ẕabṯī* crops.

Fortunately, some of the *jinsī* crops were occasionally assessed according to the *ẕabṯī* system, and therefore the *ẕabṯī* rates as well as the demand in kind and price for the same crop in the same *pargana* are available. It is consequently possible to convert the each demand per *bīgha* into demand in kind per *bīgha* (on the basis of the prevailing prices). It is also possible, by assuming that *ẕabṯī* rates were calculated on the basis of one-third of the average produce, to work out the average yield per *bīgha*. By examining the average yield per *bīgha* and the average demand in kind per *bīgha* it is possible to work out the area under each crop wherever the State demand in kind is given. Although this method is not entirely infallible because it is based on the assumption that the yields obtained in *ẕabṯī* land were the same as in *jinsī*, it provides a possible way of estimating the total number of *bīghas* under each crop.

By this method the proportion of area under different crops has been estimated (Table III). The size of the area occupied by a particular crop would, of course, be different from the share it had in total land revenue ; but trends of change in both seem to broadly synchronise.

Kharīf Crops

Bajra, *jowar*, *pulses*, *moth*, *urd* and *mung* were the most important *jinsī* crops; while *chola*, sugarcane, cotton and *maka* were the crops occupying larger areas or accounting for larger shares in revenue. In Malarna, Bahatri and Chatsu, *kodon* was also a very important crop. The other significant crops were *til* (*jinsī*), vegetables, indigo, *barti*, *madhwa*, tobacco and paddy (*ẕabṯī*).

Table III

Table showing the Percentage of Area under Different Crops

Kharīf

Pargana Jaipur

Year	*Bajra*	*Jowar*	*Moth*	*Chola*	*Sugarcane*	*Cotton*	*Maka*
1677	25	1.5	64	0.4	0.3	6.2	0.35
1744	5	4.7	41	9.1	2.3	4.8	10.7

Pargana Malarna

Year	*Bajra*	*Jowar*	*Moth*	*Kodon*	*Sugarcane*	*Cotton*	*Maka*
1690	12.6	16.6	50	—	4	4	—
1749	19.8	18	25	0.78	1.8	2.1	1.7

Pargana Bahatri

Year	*Bajra*	*Jowar*	*Moth*	*Urd*	*Kodon*	*Chola*	*Sugar-cane*	*Cotton*	*Maka*
1665	20	14	45	—	—	—	—	9	—
1725	5.7	2.6	36	20	1.2	1.3	4	14	5

Pargana Chatsu

Year	*Bajra*	*Jowar*	*Moth*	*Kodon*	*Chola*	*Sugar-cane*	*Cotton*	*Maka*
1708	55	0.5	0.5	—	1.8	—	0.3	—
1710	32	3.7	51	2	0.7	—	3	0.9
1749	10.9	2.6	33	5.5	4	0.7	18.5	19.7

Rabīʻ

Pargana Jaipur

Year	*Wheat*	*Barley*	*Gram*	*Chomli*
1	2	3	4	5
1677	3	55	40	—
1744	6.5	79	5	—

Contd

Pargana Malarna

1	2	3	4	5
1690	32	34	25	—
1749	27	42	25	—

Pargana Bahatri

1665	9	71	16	—
1725	12	52	19	—

Pargana Chatsu

1708	8	35	2	44
1710	8	55	28	2.6
1749	10	70	6	8

Rabī' Crops

Barley, wheat and gram, assessed according to the *baṭā'ī-jinsī* system were the principal *rabī'* crops; the crops assessed under *ẓabṭ* were only of marginal significance. Mixed crops such as *baijhri*, *gojai* and *gochani* were also of some significance. Another *jinsī* crop was *sarso*, but its cultivation appears to have been fairly restricted. Among *rabī' ẓabṭī* crops mention may be made of vegetables, cheena (*arzan*), tobacco and *ajwain*. But none of these yielded a significant share of revenue. In *pargana* Chatsu, *chomli* was an important crop.

The most marked change in the pattern of agricultural production one can identify was an increase in the *ẓabṭī* crops of the *kharīf* harvest presumably because the rates in general remained uniform. These were mostly cash crops. The increase is most noticeable in the following crops:

> *Varh* (sugarcane)—in *pargana* Jaipur and Bahatri; *van* (cotton)—in Malarna and Chatsu; *maka* (maize)—in Jaipur, Bahatri and Chatsu; *chola* and *kodon*—in all the *parganas*.

There was also some increase in the importance of crops like barley, wheat and gram of *rabī'* as judged from their share in revenue. The cultivation of *rabī'* crops in Eastern Rajasthan, especially wheat and barley required far greater inputs than *kharīf* crops like *bajra* and *jowar*. There was a proportionate increase in the share of the *rabī'* crop out of the total land revenue at the expense of *kharīf* cropping (Table IV). While the latter prospers reasonably well without artificial irrigation, the former requires better ploughing, watering, weeding, etc. The same is true of crops like sugarcane where better irrigation and higher inputs are essential. On this hypothesis, it would seem that during our period there was no decline in agriculture and that adequate resources were available in the village for more intensive cultivation as well as for extending cultivation.[18]

There is a corresponding decrease in the crop percentage of the *bajra* and the pulses. The increase in cash crops in *kharīf* (Table IV), accompanied as it was with a general rise in prices and a comparatively stable cash demand per *bīgha*, was not without significance for the economic life of the region. It would seem to reflect a comparative decline in local self-sufficiency, a greater use of money and an extension of commodity production.

The table on page 63 gives the number of cattle per *asāmī* in eighteen villages, yielding an average of 3.31 bullocks per peasant. Omitting two villages (one very small, the other ruined), the average in the individual villages ranged from 2.17 to 4.50 (except one village, where it is 6).[19] Obviously, the average number of bullocks with each cultivating peasant was about three, which is a fairly high rate; but this accords with other data indicating a relatively large cattle population in Mughal India.[20]

In any appraisal of agricultural efficiency, one important element to consider would be cattle, directly supplying, in the

18. Satish Chandra, 'Some Institutional Factors in Providing Capital Imputs for the Improvement and Expansion of Cultivation in Medieval India', *IHR*, July 1976, Vol. III, No. 1.
19. This table is prepared from the documents entitled *Raqbo Pargano* Mauzabad, *VS* 1698/1641. Cf. S.P. Gupta, 'Agricultural and Agrarian Society, Mid 17th Century Information from Records of *Pargana* Mauzabad in Rajasthan.' *PIHC*, Bodhgaya, 1981, pp. 370-76.
20. Cf. Irfan Habib, *Agrarian System*, pp. 53-54, and *n*.

Table IV

Proportion of Revenue from Kharīf and Rabī' Harvests

Pargana Jaipur

Year	% of Total Revenue		% of Kharīf Revenue		% of Rabī' Revenue	
	Kharīf	*Rabī'*	*Ẕabṭī*	*Jinsī*	*Ẕabṭi*	*Jinsī*
1	2	3	4	5	6	7
1677	49	51	24	76	1	99
1688	18	82	41	59	2	98
1689	38	62	32	68	1	99
1690	46	54	37	63	3	97
1712	42	58	29	71	1	99
1715	47	53	58	42	2	98
1716	33	67	62	38	3	97
1718	55	45	49	51	4	96
1720	37	63	55	45	2	98
1723	48	52	50	50	2	98
1724	89	11	60	40	30	70
1725	19	81	76	24	1	99
1733	34	66	72	28	1	99
1734	38	62	58	42	2	98
1737	39	61	57	43	1	99
1744	30	70	63	37	2	98

Pargana Malarna

Year	Kharīf	Rabī'	Ẕabṭī	Jinsī	Ẕabṭi	Jinsī
1690	70	30	29	71	2	98
1711	50	50	50	50	2	98
1712	96	4	92	8	1	99
1713	63	37	11	89	60	40
1714	56	44	23	77	—	100
1715	55	45	24	76	—	100
1716	83	17	27	73	—	100
1717	20	80	15	85	—	100
1718	79	21	23	77	—	100

Contd.

1	2	3	4	5	6	7
1719	30	70	13	87	—	100
1720	70	30	17	83	—	100
1722	81	19	13	87	—	100
1723	86	14	13	87	—	100
1726	72	28	34	66	5	95
1727	70	30	22	78	—	100
1728	79	21	18	82	5	95
1729	58	42	26	74	5	95
1732	62	38	50	50	5	95
1736	59	41	45	55	5	95
1737	52	48	31	69	5	95
1738	33	67	15	85	5	95
1743	36	64	23	77	5	95
1744	31	69	27	73	5	95
1745	61	39	20	80	—	100
Pargana Bahatri						
1665	73.4	26.6	27.8	27.2	3	97
1669	63.4	36.6	30	70	2	98
1684	48.1	51.8	40	60	2	98
1685	63.5	36.5	57	43	—	—
1686	64.8	35.2	47	53	4	96
1688	35.8	64.2	56	44	8	92
1689	53.9	46.1	60	40	8	92
1691	55.3	44.7	52	48	4	96
1696	24.2	75.8	73	27	96	4
1706	66.4	33.6	32	68	4	96
1708	74.1	25.9	31	69	4	96
1710	48.1	51.9	28	72	35	65
1716	23.7	76.3	58	42	2	98
1717	21.4	78.6	73	27	8	92
1718	71.5	28.5	39	61	7	93
1720	41.2	58.8	59	41	3	97
1721	49.1	50.9	42	58	3	97
1723	65.7	34.3	45	55	4	96
1724	55.2	44.8	36	64	3	97
1725	44.3	55.7	54	46	3	97

Contd.

Pargana Chatsu

1	2	3	4	5	6	7
1708	82	18	66	34	41	59
1709	53	47	19	81	68	32
1710	53	47	13	87	40	60
1711	55	45	20	80	6	94
1712	52	48	19	81	10	90
1713	66	34	27	73	3	97
1714	58	42	33	67	10	90
1716	54	46	42	58	13	87
1717	22	78	27	73	9	91
1721	63	37	26	74	4	96
1723	74	26	37	63	7	93
1724	58	42	26	74	6	94
1726	46	54	61	39	2	98
1730	33	67	66	34	—	100
1731	44	56	91	9	11	89
1733	53	47	49	51	13	87
1735	51	49	14	86	4	96
1737	39	61	51	49	4	96
1744	26	74	33	67	7	93
1748	45	55	52	48	12	88
1749	37	63	57	43	11	89

form of milk and milk products, a part of the rural product, and representing in draught power an important 'capital' resource of agricultural production.

In this region at least capital was available to some extent for developing agricultural production. It is, however, another matter whether the 'capital resources' were dispersed among all sections of the rural population or were concentrated in the hands of upper strata and a class of usurers. As we shall see in Chapter IV the latter is more likely to have been the case, and agricultural progress seems to have implied a corresponding growth in the dominance of the richer elements in the countryside.

Asāmīs and Bullocks in Selected Villages of Pargana Mauzabad, 1666 A.D.

Name of the village	*Asāmī*	*Bullocks*	*No. of bullocks per Asāmī*	*Castes of Asāmi owners*
1	2	3	4	5
Qasba Mauzabad	270	772	2.85	*Paṭel, Oswal, Mālī, Gujar, Khati, Telī, Chauhan, Minas etc.*
Dhamana	63	255	4.04	*Paṭel, Mālī, and misc.*
Kisanpur	13	50	3.84	*Paṭel*—30, *Jāt*—5, *misc.* 15
Gidhani	14	36	2.57	*Brahman, Gujar, Bania, Kachhwaha etc.*
Jaisinghpura	7	26	3.71	*Jāt* and *Rathore*
Sahpur	17	76	4.47	*Paṭel, Ghitala, Khati, Brahman, Jāt* and misc.
Urseva	30	180	6.00	*Dogiwal, Telī, Nirv, Ghalod, Pawar, Gaur, Naee* and misc.
Thehura	20	39	1.95	*Ujār, Paṭel*—2, *Brahman*—14, *Karwal*—2, *Naee*—8, *Jori*—6, *Jāt*—1, misc. 6.
Fagui	278	952	3.42	*Paṭel, Jāt, Sunar, Nilgar, Turak, Kheta, Paṭwārī, Ahir, Kumbhar*
Chounu	250	887	3.54	*Paṭel, Paṭwārī, Kumbhar, Khati, Telī, Brahman*
Chakwara	100	364	3.64	*Paṭwārī, Naee, Khati,* majority of the misc. belong to those whose caste is not given
Pachala	30	96	3.20	*Jāt*—33, *Paṭel*—12, *Brahman*—17, *Khati*—4, and misc.
Goindee	92	267	2.90	*Paṭel*—36, *Ahir*—53, misc. comprised of Mālī, *Gujar, Jāt,*

Contd.

1	2	3	4	5
				Paṭwārī, *Telī*, *Kumbhar*, etc.
Rampur Goindee	23	50	2.17	*Jāt*, *Paīel and* misc.
Datuli	48	155	3.22	*Paṭel*, *Jāt*, *Aheer*, *Turak* (6), *Paṭwārī*
Kagyo	—	—	—	This village was desolated and was cultivated by the *rai'yat* of an adjoining village named Datuli
Jhak	116	320	2.75	*Paṭel*, *Paṭwārī*, *Jāt*, *Khati*, *Brahman*, *Aheer*, *Naee*, *Kumbhar*, *Gujar* and misc.
Khatwar	69	205	2.97	*Paṭel*—21, *Brahman*—35, *Jāt*—15, *Chaudhurī*—2, *Gujar*—5, *Mina*—6, *Paṭwārī*—7, misc.—116.
Kurhaz	17	40	2.35	*Paṭel*—8, *Pa'i*—3, *Gujar*—14, & misc.—15.

Appendix to Chapter II

Table (a)

Pargana Bahatri[21] *VS. 1797/A.D. 1740*

Crop	*Area (bīghas)*	*Praduce (in mans per bīgha)*	*Total Production (in mans)*	*State share (in mans)*	*Prices (in mans per rupee)*	*Total Amount (Cash)*
Barley	16,202	8. 0.0	129,616	43,205.13.0	2. 0.0	21,602.11.0
Wheat	4,555	7. 0.0	31,885	10,628.14.0	1.10.0	8,503. 2.1
Gram	10,042	2.20.0	25,105	8,368.14.0	1.20.0	5,578.14.0

21. *Takhmīna sakh unahlu (rabī'), pargana* Bahatri *VS* 1797/1740.

Table (b)

Bahatri VS 1800/1743[22]

Mauẓa	*Revenue*	*Māl*	*Vor Kyari*[23] *bīgha*	*Amount*	*Crop*	*Area*	*Yield per bīgha*	*Total produce*	*One-third of total produce as share of state*[24]	*Price in mnas. per Rupee*	*Amount*
	Rs.	*Rs.*		*Rs.*			*(mans)*	*(mans)*	*(mans)*		*Rs.*
1	2	3	4	5	6	7	8	9	10	11	12
Qaṣba Baswa	5120	4780	3200 (*kyari*)	4720	Barley	2400	7.00	16800	4772	1.25.0	2980
			1000 (Vor)		Wheat	800	7.00	5600	1800	1. 5.0	1540
Karoth	514	464.0.0	414	200	Barley	200	7.00	1400	466	1.25.0	300
					Wheat	50	7.00	350	117	1. 5.0	104
Sakat	2385	2135	1500	2110	Barley	1000	7.00	7000	2000	1.25.0	1225
					Wheat	500	7.00	3500	1000	1. 5.0	815
Goth *Kalān*	221	198	110 (*kyari*)	176	Barley	80	7.00	560	187	1.25.0	114
			12 (Vor)		Wheat	30	7.00	210	70	1.5. 0	62

22. *Jamadol sakh Unahlu* (*Rabī'*) *VS* 1800/1743.
23. *Kyari* and *Vor* denote the quality of land based on the means of cultivation.
24. These figures are not accurate in calculations. I have given them here as they are in the original documents.

Contd.

Niwla	1110	1010	650	985	Barley	450	7.00	3150	1000	1.25.0	610
					Wheat	150	7.00	1150	350	1. 5.0	375
Motipara	1105	1005	905 (*kyari*)		Barley	400	7.00	2800	900	1.25.0	575
			10 (*Vor*)		Wheat	150	7.00	1050	350	1. 5.0	330
Bishanpura	101	96	91		Barley	64	7.00	450	150	1.25.0	80
					Wheat	50	7.00	35	12	1. 5.0	11
Budhsinghpura	26.8.0	24.12.0			Barley	10	7.00	70	23	1.25.0	14.4.0
					Wheat	5	7.00	35	12	1. 5.0	10.8.0
Narharwas	111	100	80 (*kyari bīgha*)		Barley	10	7.00	70	24	1.25.0	15.0.0
					Wheat	30	7.00	210	70	1.5 .0	65.0.0

Table (c)

(i) *Tappa Muhi*

Mauza	*Crop*	*Area*	*Production per bīgha (mans)*	*Total Produce (mans)*	*State Share (One-third of total produce) (mans)*	*Prices in mans. per rupee*	*Amount (in rupees)*
Gadha	Barley	350	7	2450	816.20.0	0.25.0	510.0.0
	Wheat	50	7		116.20.0	1. 5.0	110.0.0
	Gram	70					25.0.0
Sawatheri	Barley	50	7	350	116.27	1.25.0	70.0.0
Brahmin Para	Barley	400	7	2800	933.14.0	1.25.0	575.0.0
	Wheat	40	7	280	93.13.0	1. 5.0	85.0.0
Reta	Barley	200	7	1400	466.27.0	1.25.0	290.0.0
	Wheat	20	7	140	46.27.0	1. 5.0	45.0.0
	Gram	100		—		—	25 0.0
Abari	Barley	150	7	1050	350	1.25.0	215.0.0
	Wheat	40	7	280	93.14.0	1. 5.0	85.0.0
	Gram	100	—	—	—	—	25.0.0
Mitrapari	Barley	50	7	350	116.20 0	1.25.0	72.3.2

Table (d)

(ii) Tappa Kolesar

Mauẓa	*Crop*	*Area*	*Production per bīgha (mans)*	*Total Produce (mans)*	*State Share (One-third of total Produce) (mans)*	*Prices (mans per Re.)*	*Amout of Revenue (Rs.)*
Kolesar *Kalān*	Barley	100	7	700	233.13.0	1.25.0	144. 8.0
	Wheat	25	7	175	85.13.0	1. 5.0	51.12.0
Khohra *Khurd*	Barley	150	7	1050	350	1.25.0	215. 0.0
	Wheat	50	7	350	116.27.0	1. 5.0	103.12.0
Paṇpura	Barley	80	7	560	186 27.0	1.25.0	114. 0.0
	Wheat	20	7	140	46.27.0	1. 5.0	41. 8.0
Siṅghpur	Barley	50	7	350	116.27.0	1.25.0	72. 0.0
	Wheat	20	7	140	46.27.0	1. 5 0	41. 8.0
Nawagaon	Barley	100	7	700	233.14.0	1.25 0	143.12.0
	Wheat	20	7	140	46.20.0	1. 5.0	41. 8.0

CHAPTER III

Agricultural Prices

Quotations of the agricultural prices come mainly from the *arhsattas*, which give either the actual sale price of grain collected as revenue or the notional price of the revenue realized in grain.[1] The prices of crops (chiefly chola, kodon, maize, sugarcane and cotton), for which revenue was charged by measurement (*ẓabṭī*) are so rarely given as to be inadequate for comparative analysis. The important food-grains for which prices are consistently available (under the record of the *jinsī* collections) are :

In *Kharīf* : Bajra, Jowar, Moth, Urd and Mung.
In *Rabī'* : Wheat, Barley and Gram.

In our study, all the original quotations and implied prices have been converted into rupees (fractions being calculated in decimals) per *man* ;[2] in the original, the figures are given in quantity per rupee.[3] We have constantly to bear

1. See. S. Nurul Hasan and Satya Prakash Gupta "Price of Food-Grains in the Territories of Amber", *PIHC*, Patiala, 1967, pp. 345-368.
2. In the documents the price is stated in rupees. For smaller fractions *ṭakas* (sometimes also referred as *murādi*) are used; the number of *ṭakas* in a rupee was not fixed but changed from year to year and *pargana* to *pargana* (see Table 6 in S. Nurul Hasan & S.P. Gupta, 'Price of Food Grains in the Territories of Amber', *op. cit.* Each *ṭaka* was further divided into 50 *dāms*, which were minute fractional units of account, quite different from the *dāms* coined by Akbar and his successors.
3. *Nirakh bāzār* (Pers. *nirkh-i-bāzār* of *pargana* Amber, Chatsu, Dausa, Lalsot, Bahatri and Malarna. See also *arhsattas* of different *parganas*.

in mind the fluctuations in the number of *sers* per *man* from one *pargana* to another. The Akbarī *man* survived curiously enough as a *man* of 30 *sers* (and not, as originally, as a *man* of 40 *sers,* each *ser* of 30 *dām*-weights) ; similarly the Jahāngīrī *man* was preserved as a *man* of '36 *sers*' ; the standard *man* (Shāhjahānī) was of '40 *sers*'.[4] This thus presumes a standard *ser* of Shāhjahānī weight in which the Akbari and Jahāngīrī *man* now came to be expressed. We have converted all other *man* into Shāhjahānī *man* for convenience of comparison.

One may assume that the prices given in the *arhsattas* are harvest prices. It is, however, possible that some sale-prices reflect post-harvest rates. Some quotations might also reflect the varying rates for different qualities of the same grain.

The prices of major crops obtained from the *arhsattas* for a select group of contiguous *parganas* are set out in Figures 1-7, the lower curve representing the lowest rates among the *parganas* in each year, and the upper the highest. An examination of these curves reveals considerable variations in prices prevailing in different *parganas*. The variation is less marked in the case of *rabī'* crops than in the *kharīf* crops. Further, the variation tends to be less during the reign of Aurangzeb than in the period 1710-1750. It may also be observed that the relative values of the different food-grains (say, with wheat as base) differs from *pargana* to *pargana* and the changes in the values of the different grains over the years do not follow a uniform pattern either. (See Table 1.)

It will be seen that the *rabī'* crops maintain steadier relationship with wheat over the years than do the *kharīf* crops. But this was to be expected, since wheat itself shared in the fortunes of the *rabi'* harvest, while the *kharīf* harvest could have been of a completely different quality from that of the *rabī'* harvest. But in the long term the relative prices present a cogent enough picture.

The differences in levels of prices of foodgrains among *parganas* may partly be explained by the difference in productivity

4. The number of *sers* going to a *man* in Amber and Chatsu was 40, in Dausa and Lalsot 30, and in Bahatri and Malarna, 28.

PRICES OF MAJOR CROPS (FIGS. 1 To 7)

WHEAT

FIG. 1

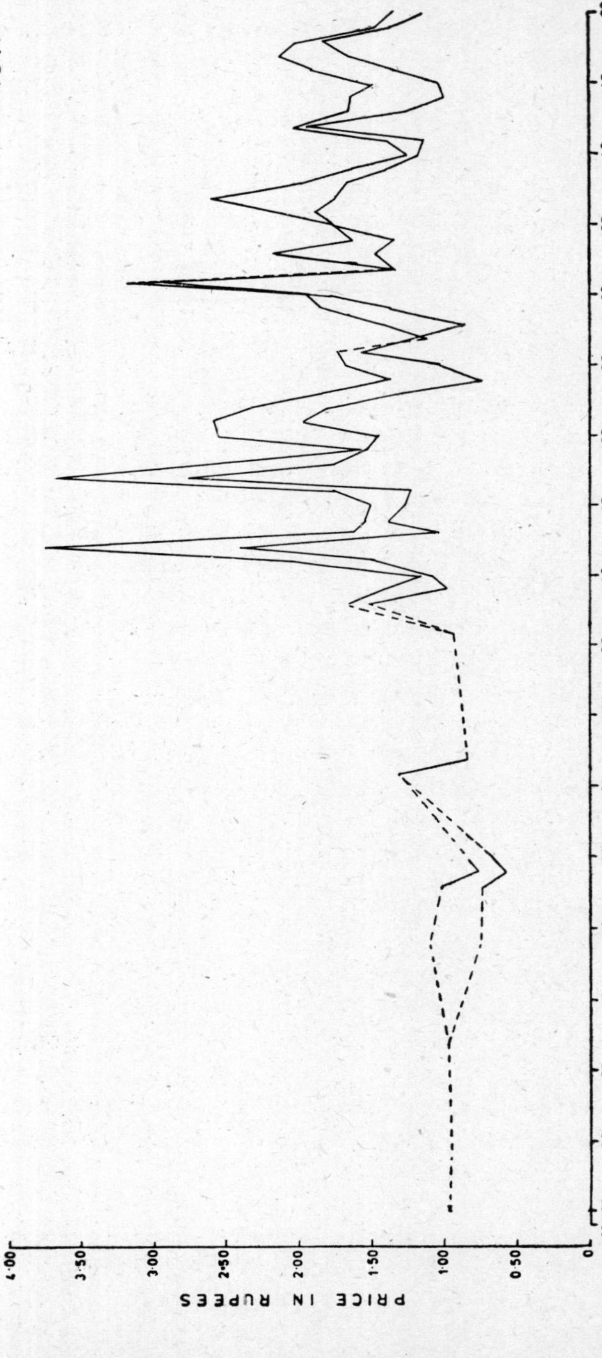

FIG.2

PRICES OF MAJOR CROPS (FIGS. 1 To 7)

GRAM

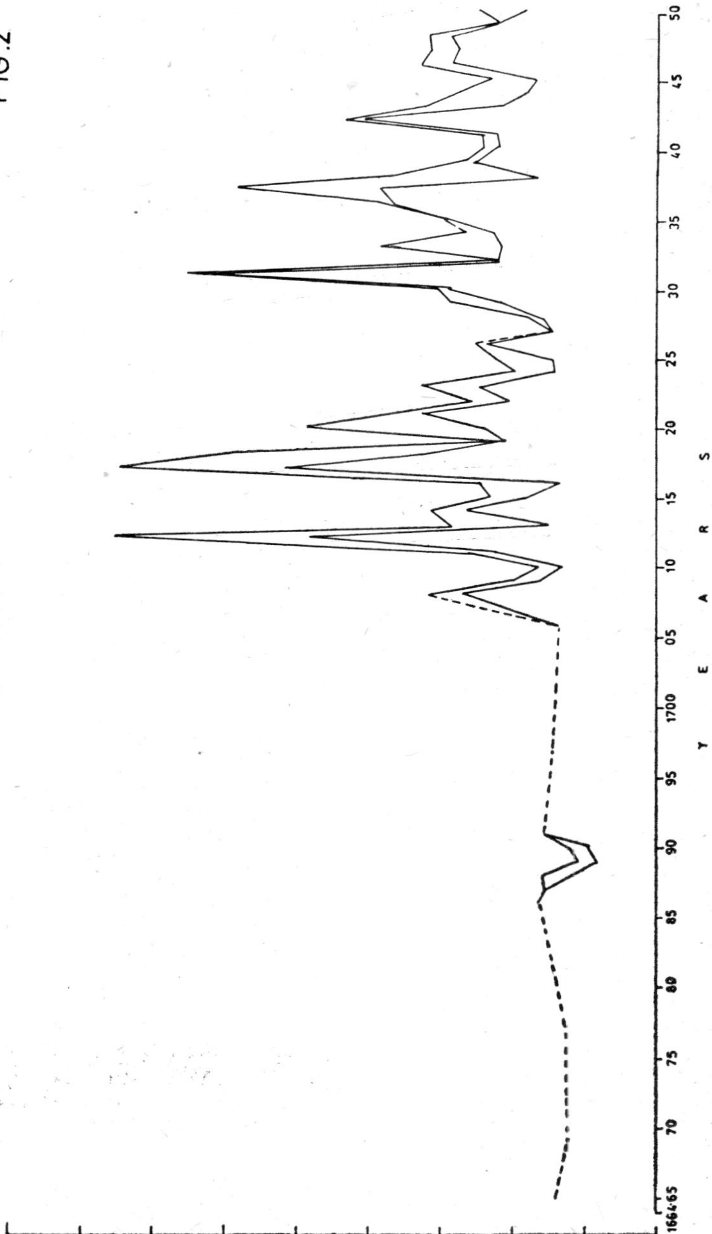

PRICES OF MAJOR CROPS (FIGS. 1 To 7)

BARLEY

FIG. 3

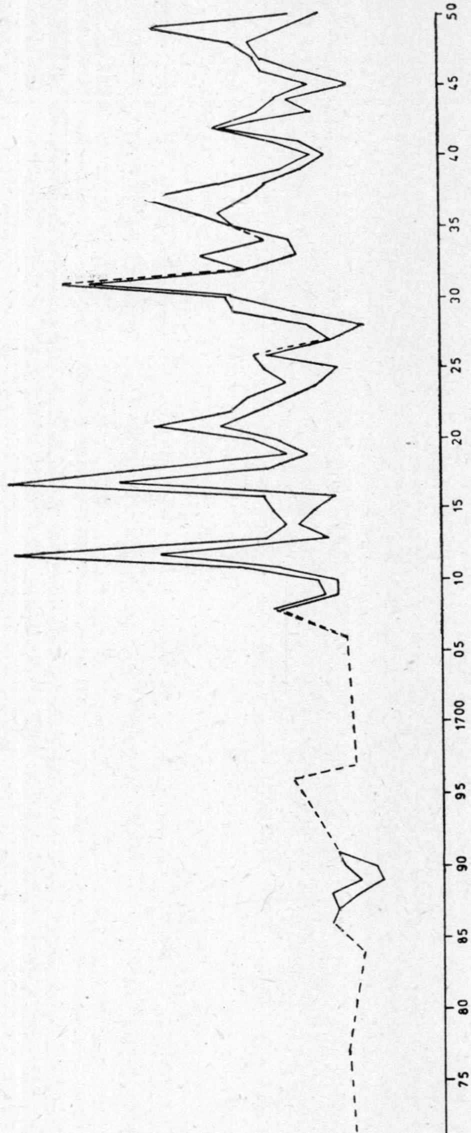

PRICES OF MAJOR CROPS (FIGS. 1 To 7)

BAJRA

FIG. 4

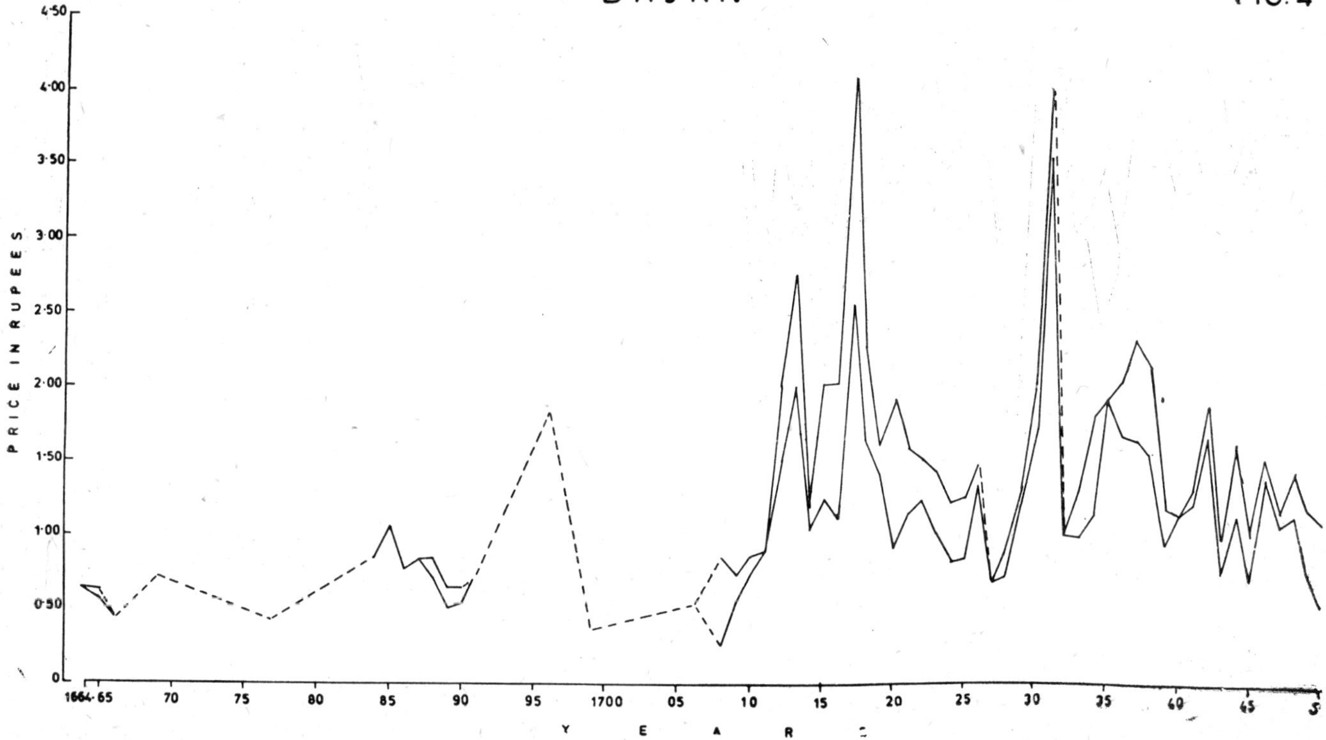

FIG. 5

PRICES OF MAJOR CROPS (FIGS. 1 To 7)

JOWAR

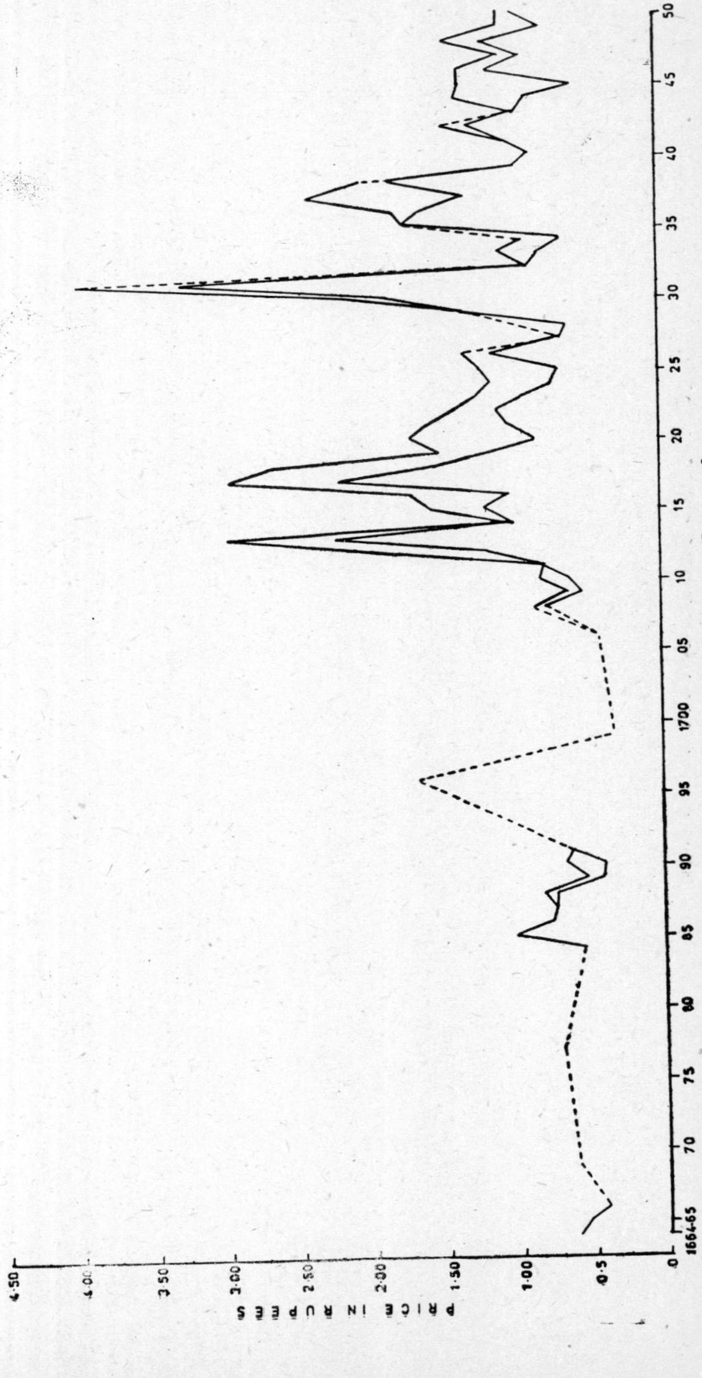

FIG. 6

PRICES OF MAJOR CROPS (FIGS. 1 To 7)

MOTH

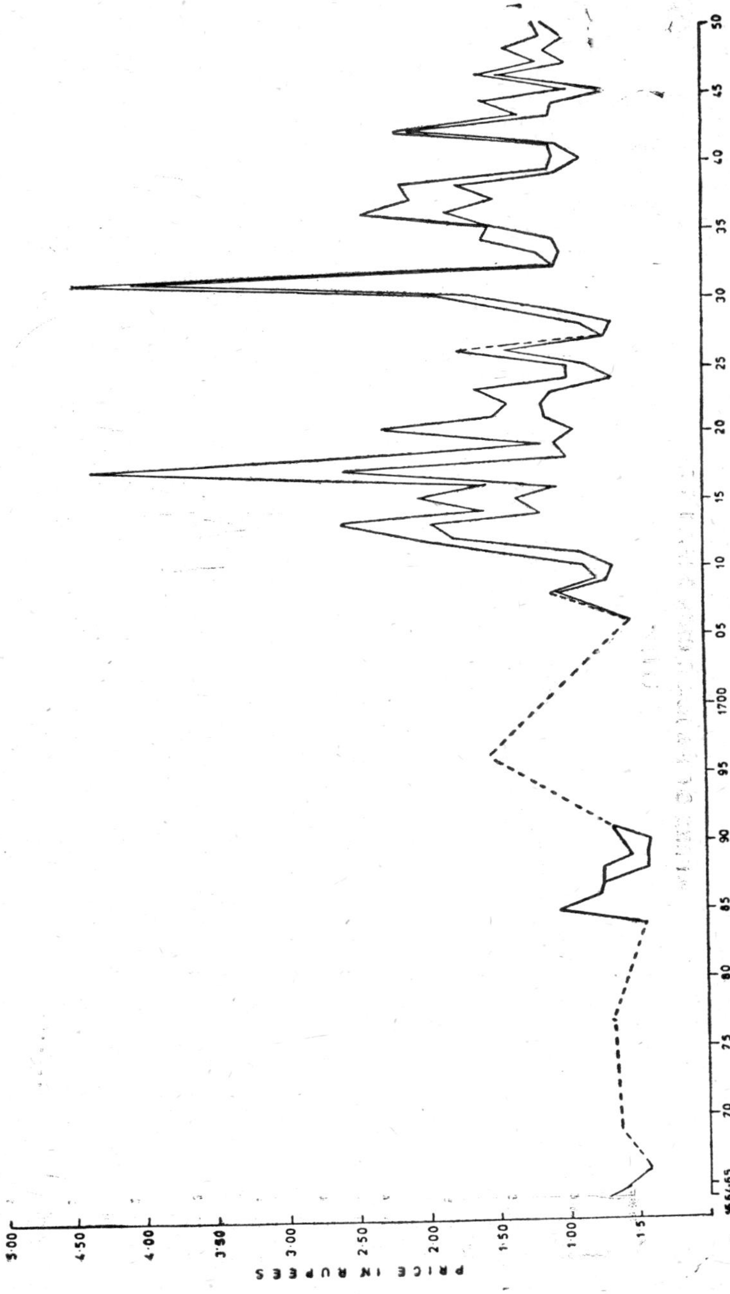

PRICES OF MAJOR CROPS (FIGS. 1 To 7)

URD

FIG. 7

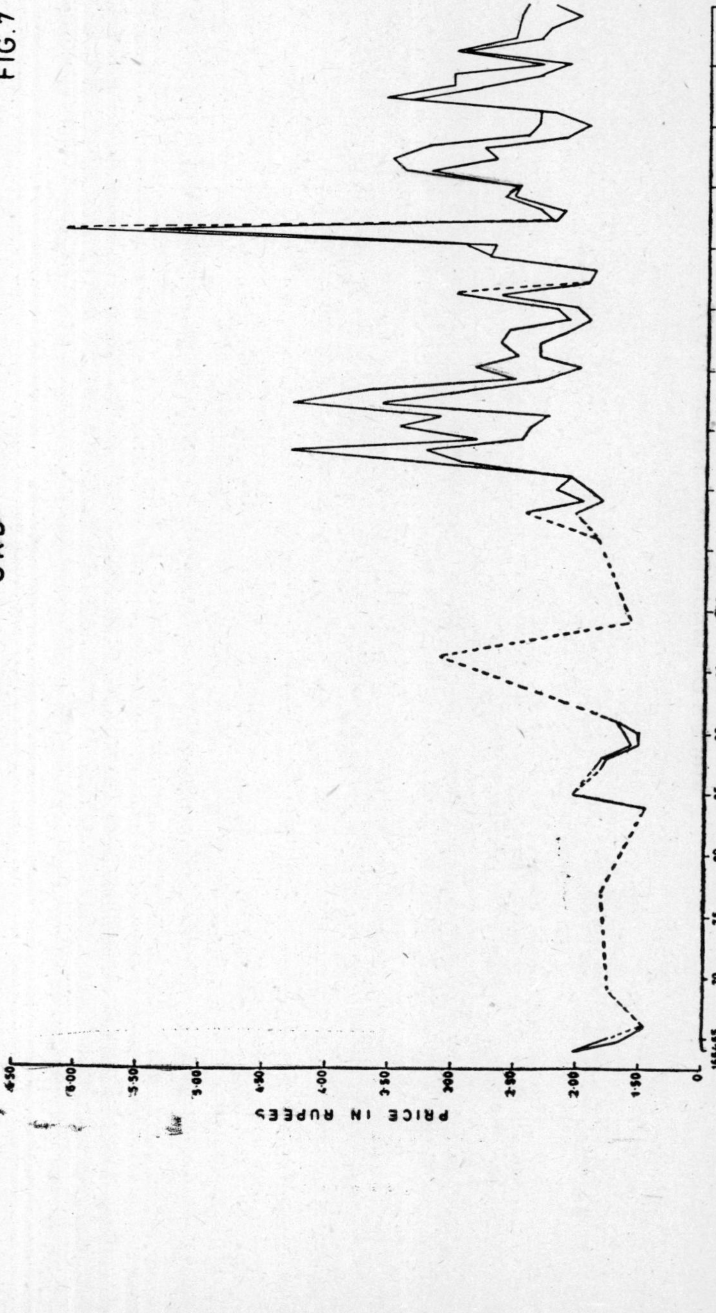

Table 1

Showing the relative prices of different food-grains

(Index : Wheat as 100 in each *pargana* in each year)

Year	Wheat	Barley	Gram	Bajra	Jowar	Moth	Urd	Mung
				Amber				
1677	100	67	67	45	74	71	84	92
1715	100	70	72	100	98	103	110	131
1725	100	67	72	85	67	80	115	118
1734	100	76	86	88	66	80	115	92
1744	100	—	94	116	88	104	131	130
				Chatsu				
1714	100	71	94	79	68	86	124	151
1726	100	74	78	87	77	100	111	133
1735	100	79	84	108	96	83	85	97
1744	100	67	85	101	84	94	126	133
				Dausa				
1709	100	72	84	58	55	69	85	88
1715	100	76	77	99	86	106	114	124
1736	100	79	104	108	95	110	118	140
1744	100	66	85	96	81	84	110	100
1750	100	74	90	72	72	86	108	93
				Lalsot				
1715	100	73	75	150	—	151	184	191
1734	100	73	82	110	100	94	100	112
1741	100	97	106	116	—	100	115	134
1745	100	58	67	58	88	57	90	90
				Bahatri				
1665	100	66	70	62	56	59	72	59
1689	100	71	70	82	67	65	72	87
1716	100	68	69	114	96	87	81	111
1725	100	70	67	75	73	57	63	68
				Malarna				
1690	100	79	73	82	81	70	75	84
1715	100	68	70	106	118	123	145	151
1726	100	76	76	88	74	89	104	100
1732	100	80	84	77	66	78	91	88
1744	100	77	87	85	71	95	144	129

and geographical conditions. Revenue rates which were partly based on yields should therefore also have varied considerably as between adjoining *parganas*. For example, during the reign of Akbar, Amber had one *dastūr* (revenue rate), while Chatsu and Malarna had another. The difference in the standard revenue per *bīgha* was considerable, as may be seen from the table below :

Table 2

Standard demand per bīgha in Dāms & Jītals[5]

Kharīf

	Amber		*Chatsu and Malarna*	
	D	*J*	*D*	*J*
Jowar	11	16	36	22
Moth	15	16	22	9
Urd	29	2	39	3
Mung	15	16	36	22

Rabī'

	Amber		*Chatsu and Malarna*	
	D	*J*	*D*	*J*
Wheat	31	8	53	18
Barley	20	3	38	0
Gram	20	3	38	0

The differences may also be attributed to a variety of other factors. The foodgrains which were consumed locally and did not enter regional or long-distance trade, tended to show greater variation. This may be noticed in the case of *jowar* and *bajra*. In the case of wheat, gram and barley, the difference in price is not as marked. Another important factor would be the increase or decrease of area under a particular crop (for which see Chapter II).[6] Finally, the inadequacy of means of transport

5. *Āin-ī Akbari*, ed. Blochmann, Vol. I, pp. 364-67.
6. See S. Nurul Hasan, K.N. Hasan & S.P. Gupta "The Pattern of Agricultural Production in the Territories of Amber', *PIHC*, Mysore, 1966.

must have certainly contributed materially to the variation in prices even in contiguous *parganas*.

The *arhsattas* also furnish us with prices prevailing in different villages in the same *pargana*. In Table 3, we offer a sampling of the range of village-level quotations from three *parganas*. It appears that the range of variation in prices in different villages was small, suggesting that the isolation of the village from the market was not as great as is sometimes thought.

Table 3

Variation in the prices of food-grains in Villages of the same Pargana

(in rupess per *man*)

	Chatsu 1664	Chatsu 1730	Malarna 1730
		Kharīf	
Bajra	0.55 to 0.68	1.66 to 2.00	2.00 to 2.31
Jowar	0.59 to 0.69	1.66 to 1.90	2.00 to 2.28
Moth	0.73 to 0.85	1.48 to 1.90	1.74 to 2.10
Urd	1.00 to 1.14	1.55 to 1.90	1.85 to 2.10
Mung	1.21 to 1.29	1.66 to 2.10	1.81 to 2.10
		Rabī'	
Wheat	NA	1.66 to 1.78	1.79 to 2.00
Barley	NA	1.29 to 1.43	1.37 to 1.50
Gram	NA	1.42 to 1.55	1.37 to 1.54

We thus get to the conclusion that the price variations in geographical terms tended to be very small within distances where transport costs did not enter materially in the determination of price. But as the distances increased, the price range became wider and wider even though the *parganas* within which the two markets lay were contiguous. In other words, Eastern Rajasthan constituted for the bulkier grains an assemblage of separate local markets, not so much because of any institutional drawbacks but essentially because of transport costs.

Moreover, even given the existence of separate markets, the impulses (of the kind we have discussed) that moved prices within them might yet be common. This can be tested by looking at the actual movements of prices over time from the graphs already made.

There was considerable fluctuation in the price of the different commodities from year to year, varying again from crop to crop. One noticeable feature is that any sharp rise as well as any decline tends to affect all or most of the *parganas*. The most unusual year of price increase is, 1712 ; after which prices never came down to the level prevailing during the reign of Aurangzeb. The other years of significant ascents are 1717-18, 1731 and 1737. In the forties the prices tend to decline. It may be pointed out that during the reign of Aurangzeb, the price of wheat generally remains under a rupee per *man* and that of the other foodgrains is considerably less. By the middle of the 18th century these prices tend to be over one rupee per *man*.

This broad picture of increase in the price level may be illustrated by the price of indigo prevailing in the *pargana* of Amber (Table 4). We take this as an index because, being a product easy to transport, indigo might be thought to reflect general price movements better than any other crop.

Table 4

Prices of Indigo in Pargana Amber

Year	*Price*	*Year*	*Price*
1666	25.28	1720	41.72
1677	23.33	1723	52.00
1688	25.98	1724	37.40
1689	28.29	1725	44.50
1690	28.58	1726	40.00
1712	38.68	1729	38.50
1715	51.66	1733	59.60
1716	45.61	1737	47.00
1718	45.30		

The other phenomenon to notice is an upward tendency, after the first quarter of the 18th century, which is also established from the movement of the rate indices ; but a decline follows the rise before the middle of the 18th century.

The graph of running decennial averages of weighted prices (all-*pargana*) which may be seen in Figure 9 brings out the results set out above in a more definite form.

But it is difficult to explain the secular trend in price movements, even when the decennial running average define them so well. One possibility is that, in part at least, these are related to prices of silver. As silver flowed into the country, its price rose and stabilised in correspondence with the ebb and flow of the silver influx. Copper which during this period was a more stable metal (in respect of its value) reflects a continuous depreciation of silver, as may be seen from our graph on Figure 10.

The movement of rupee prices of copper can be compared with Decennial Running Average of Weighted Prices (Figure 9), but only from the decade ending 1717. The comparison thus covers the first part of the 18th century. It will be seen that as the fluctuations in agricultural prices are smoothed out through constructing decennial averages, the resultant curve shows a stability which matches that of the copper index. Yet in the early years copper tends to be lower in relation to its 1715 price than agricultural prices. Within the last twenty years, on the other hand, the agricultural prices fall substantially behind copper. Clearly, while depreciaton of the silver price continued, local harvests in Eastern Rajasthan continued to exert pressure enough on prices to create significant medium-term deviations from the movement of the silver-copper ratio. Thus even if there was a general long-term increase in prices over the century (1650-1750), owing to monetary phenomena, the movements of agricultural prices over shorter periods in their ascents and descents reflected other factors as well, such as weather-cycles, extension of agricultural and—most elusive of all—trends in population growth. On these, at present, little or nothing can be said, except that price fluctuations can, from our statistics, be precisely fairly delineated, and this provides a reasonably firm ground to begin to look for correlation with the other possible factors.

RUNNING DECENNIAL AVERAGE OF WEIGHTED PRICE INDICES
(BASE YEAR 1715)

FIG. 9

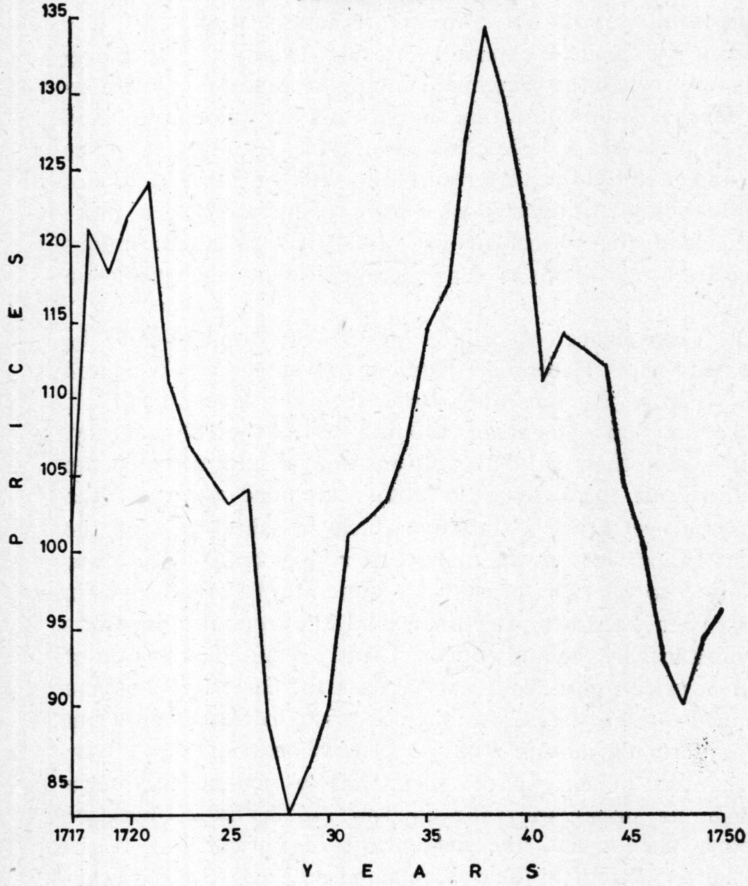

INDICES OF COPPER PRICES
(BASE YEAR 1715)

FIG. 10

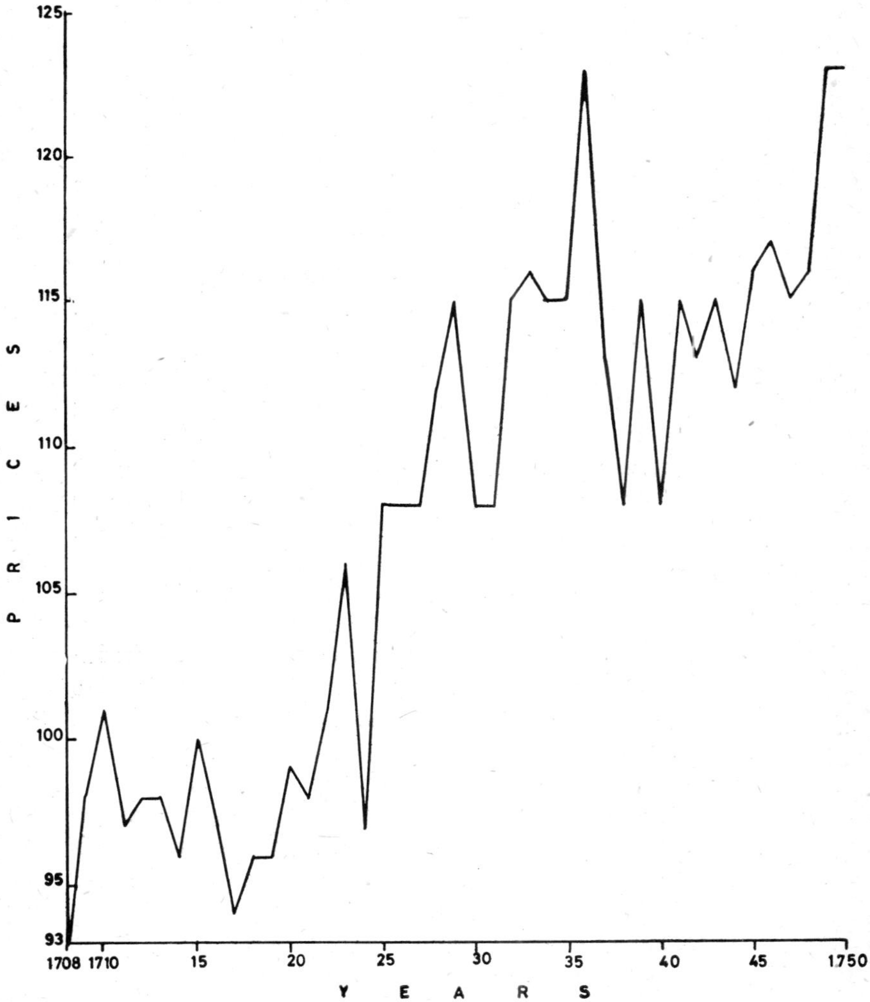

CHAPTER IV

Agrarian Trade

We have seen that the state often appropriated about one half of the produce of the cultivator as land-revenue on most of the crops. In addition, the peasant had to pay roughly a tenth of the produce directly or indirectly by way of various taxes in cash and kind. The collection of land-revenue in cash was a fairly well established practice in the whole of the Mug͟hal Empire; and this seems to have been true for Eastern Rajasthan as well (Chapter II). This naturally induced the rise of local markets, where the peasant had to sell his produce in order to make land-revenue payments.

As explained in Chapter II, two methods of revenue assessment/collection generally prevailed in Eastern Rajasthan. One was *ẓabṭī*, where the revenue demand was stated in cash per *bīgha*; the other *jinsī* or *baṭā'ī-jinsī*, where the revenue was originally fixed in kind.

Arhsatta and *Jama'bandī* documents show that *jinsī* accounted for the bulk of revenue collected in this region. Figures 1-6 give us the share of the revenue realized under *ẓabṭī* (Ẓ) and *jinsī* (J) in different *parganas*.

Out of the six *parganas* it is only in *pargana* Lalsot that *ẓabṭī* pre-dominated. This might be due to the fact that throughout our period the Mug͟hal *manṣabdārs* directly controlled this *pargana*. But even if *jinsī* predominated it did not mean that the peasant did not have to pay in money.

From the *jama'bandī* of different *parganas* of Eastern Rajasthan for the second half of the 17th century, we can also see that where the state collected the revenue in kind, its immediate impulse was to sell it (or the claims to it) to anyone ready to purchase it at the approved prices (*nirkh-qarār*).

The grain collected in payment of revenue was put on sale (*bichotī*) in the village itself at harvest prices ; the remaining portion known as *anya* in our document was sent to the *kothar* (in the village itself) and finally to *awar* (state granary) under the jurisdiction of a *awārdār* (incharge of *awār*). The grain was stored here presumably for the purpose of meeting emergencies. Such stores could be sold later at market prices. One can see from Table I that the proportion of the grain collected for store (*anya*) was very small as compared to the collection in money through village-level sale (*bichotī*). The selling prices at *bichotī* at the village level are generally found to be lower (being close to the harvest prices) than the prices gained by sale from government granaries ; but this to be expected since the latter must have been close to urban market prices.

The *mahājan* and *bania* played an important role in transporting the revenue-grain, either taking it to the nearest market or to the state granary (*awar*).[1]

Under the *bichotī* arrangements, which as we have seen, accounted for the larger part of revenue, the authorities used the agency of a *vyopari*, village *baniā* (*vana*, *vanī*) or *mahājan* of the *qaṣba*. It is evident from the *chiṭṭhīs* of different *parganas* of different years sent to the *dīwān* by *pargana* officials (*'āmil* and *amīn* in particular) that the object was to convert the collected grain into cash on the best terms. A *chiṭṭhī* of 1664 from *āmil* Akhay Raj of *pargana* Mauzabad informed the then *dīwān* Kalyan Das that "after the collection of *jinsī* as state share, he (the *'āmil*) wished to sell it at the spot. But no one was prepared to purchase it. Then the *'āmil* called a *bania* from Phagi (an adjacent *pargana* to Mauzabad, see map 1) and asked him to purchase the grain (*nāj*). The *baniā* said that he neither had the money nor the capacity to purchase the grain."[2] It thus appears that the state faced some difficulty in selling off *jinsī* owing to lack of resources with the merchants.

1. *Chiṭṭhī* from Akhay Raj to *dīwan* Kalyan Das *VS* 1721/1664, *Chiṭṭhi* of *dīwan* to Khushal Chand Kishan Chand *VS* 1800/1743 ; *Chiṭṭhī* of Kushal Singh Udai Ram and Anand Ram (*pargana* officials) to *dīwan* Bhikhari Das and Fateh Chand *VS* 1771/1714 ; *Chiṭṭhī* from Amra *Kotwal* to *dīwān* Kalyan Das *VS* 1722/1665. *Roznāma-waqia-VS* 1745/1668.
2. *Chiṭṭhī pargana* Mauzabad *VS* 1721/1664.

RELATIVE SIZE OF ASSESSMENTS IN KIND AND CASH

% of Total Revenue

PARGANA AMBER

FIG.1

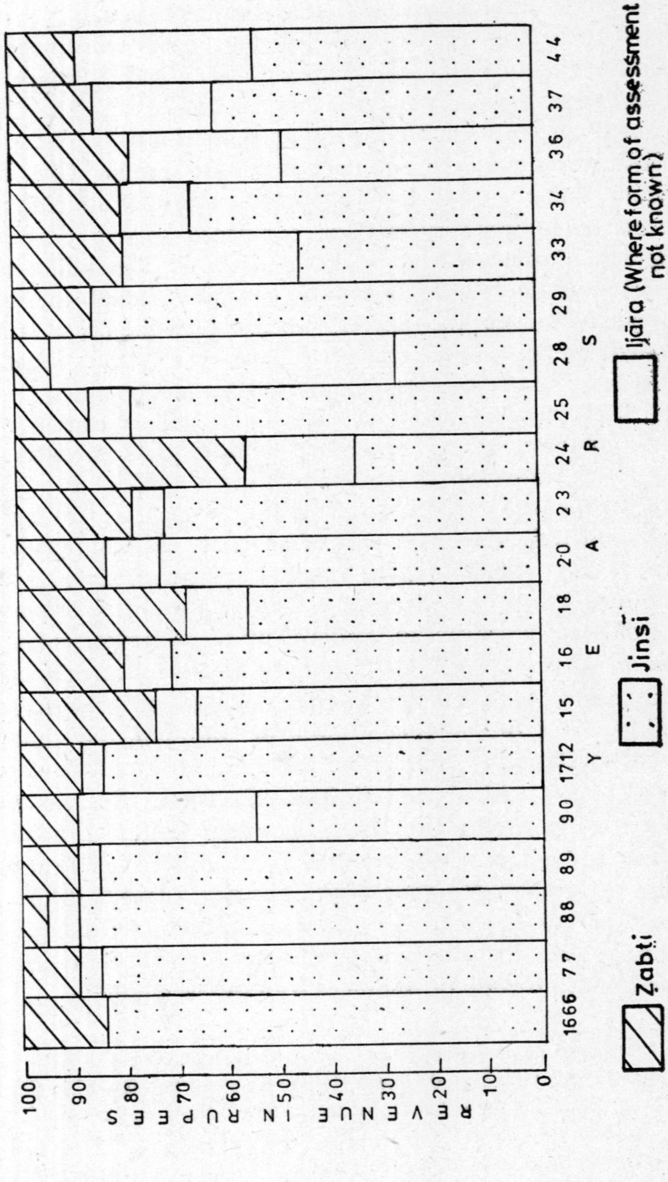

RELATIVE SIZE OF ASSESSMENTS IN KIND AND CASH

% of Total Revenue

PARGANA BAHATRI

FIG. 2

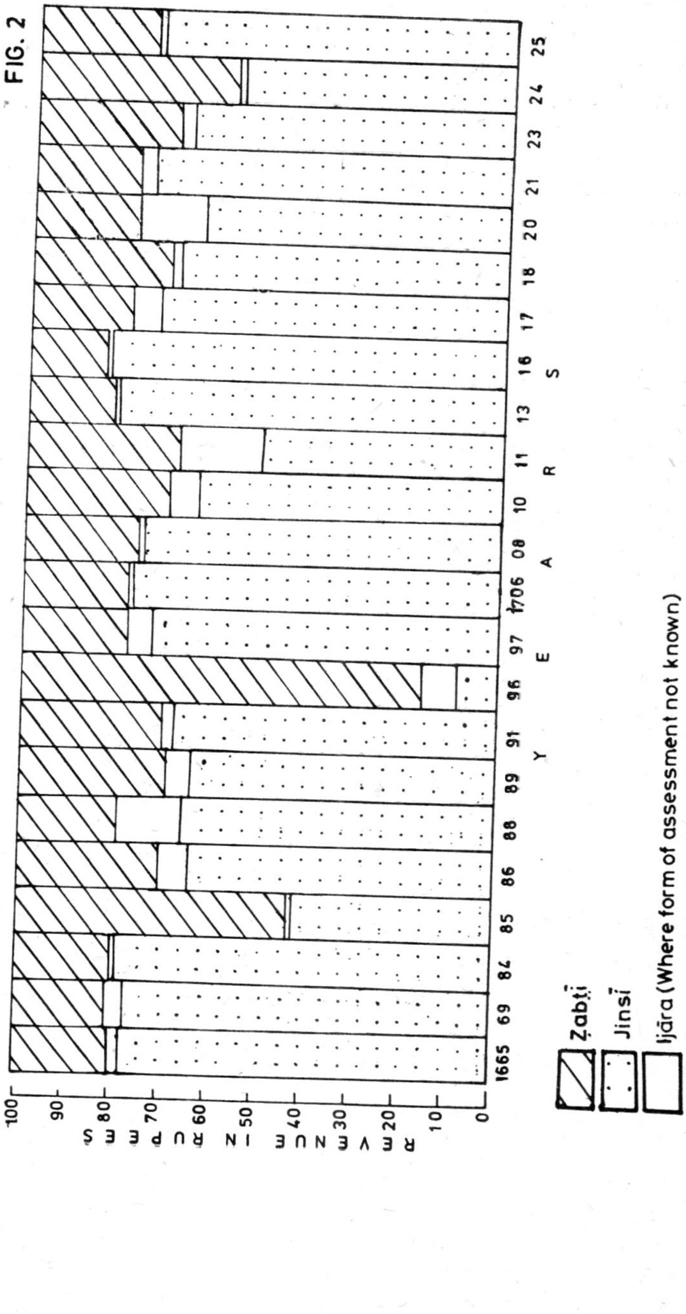

FIG. 3

RELATIVE SIZE OF ASSESSMENTS IN KIND AND CASH

% of Total Revenue

PARGANA CHATSU

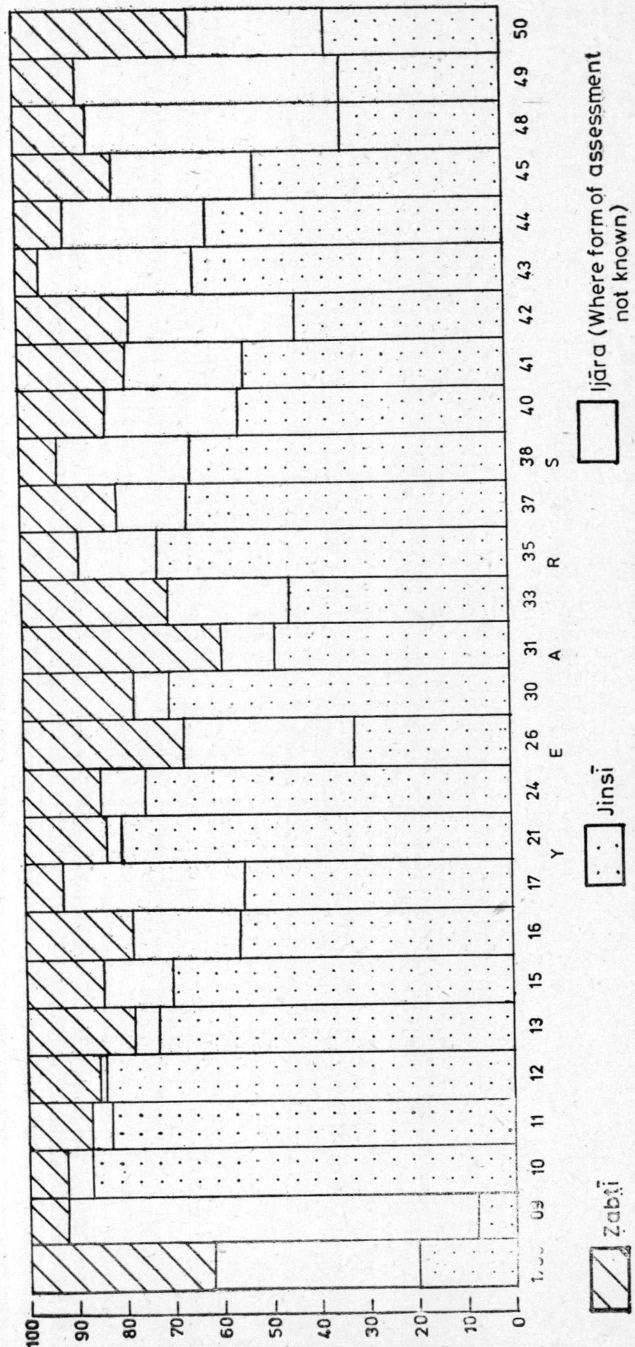

RELATIVE SIZE OF ASSESSMENTS IN KIND AND CASH

% of Total Revenue

PARGANA DAUSA

FIG. 4

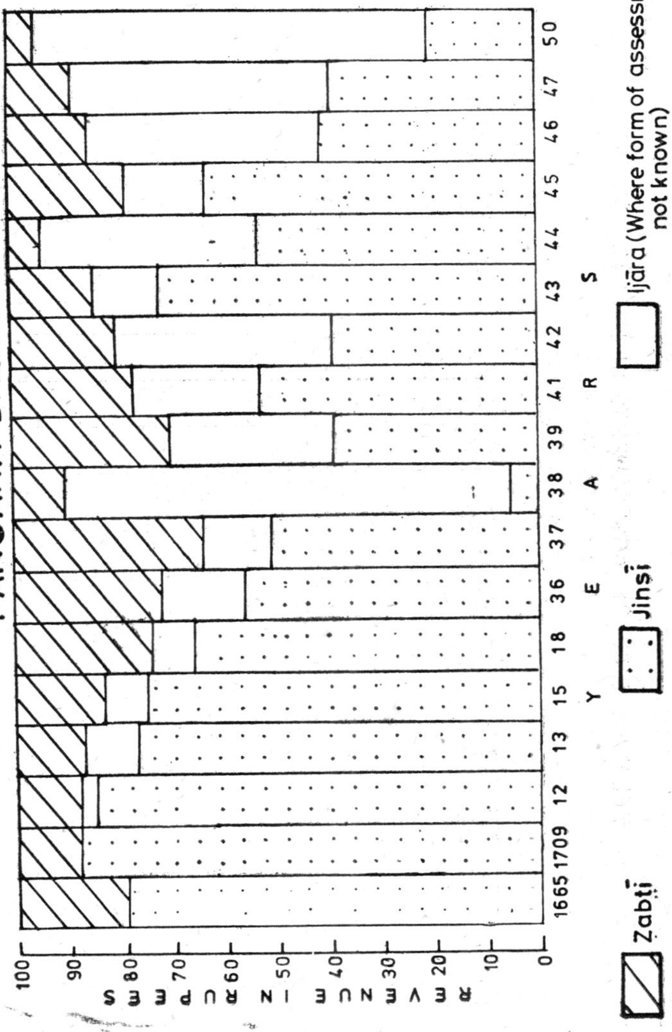

RELATIVE SIZE OF ASSESSMENTS IN KIND AND CASH

% of Total Revenue

PARGANA LALSOT

FIG. 5

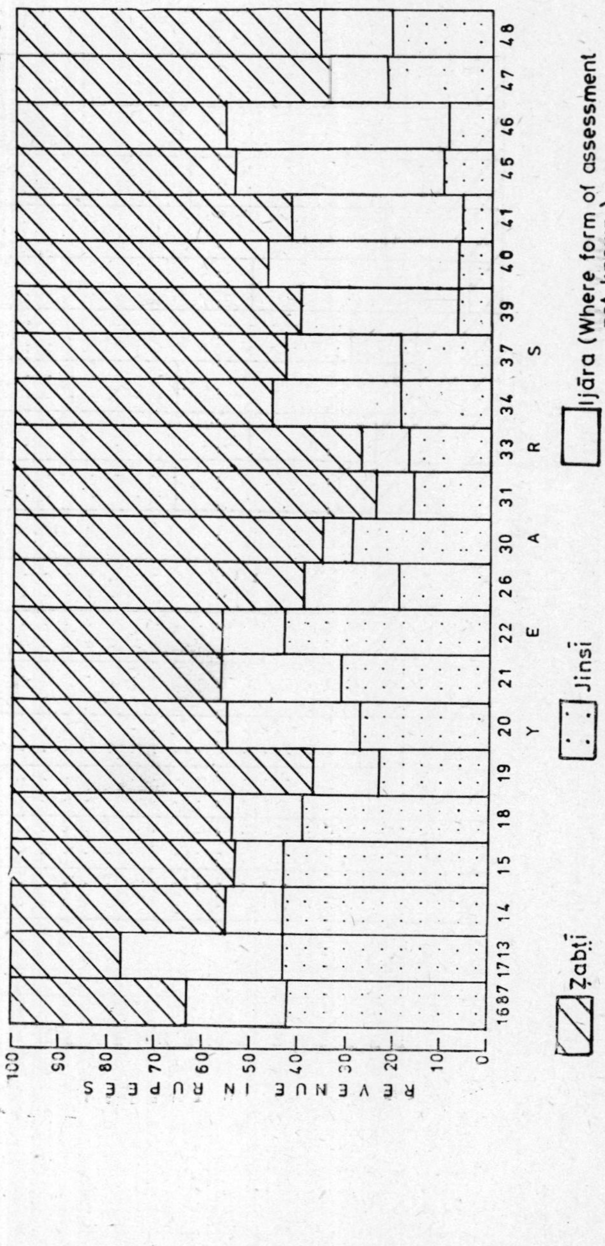

FIG. 6

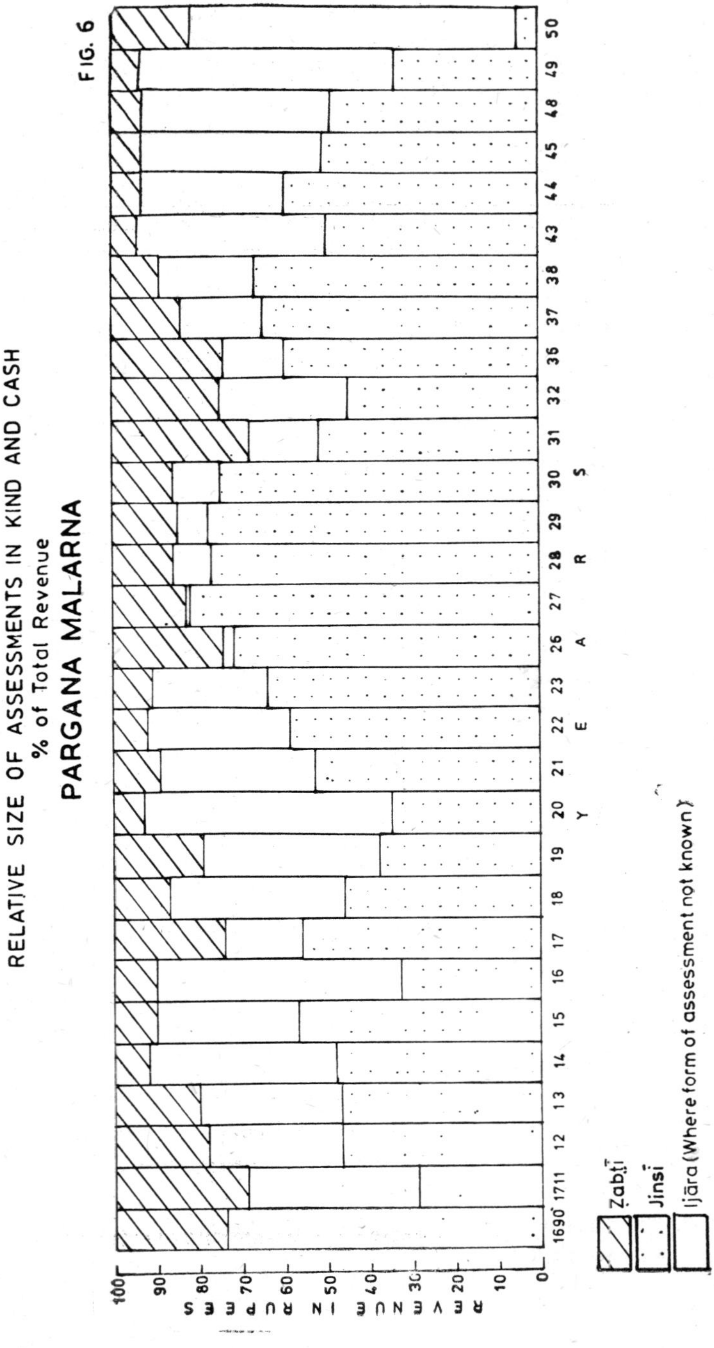
RELATIVE SIZE OF ASSESSMENTS IN KIND AND CASH
% of Total Revenue
PARGANA MALARNA
REVENUE IN RUPEES
100
90
80
70
60
50
40
30
20
10
0
1690
1711
12
13
14
15
16
17
18
19
20
21
22
23
26
27
28
29
30
31
32
36
37
38
43
44
45
48
49
50
YEARS
Ẓabṭī
Jinsī
Ijāra (Where form of assessment not known)

Table I[3]

Percentage of Jinsi commuted into cash (Bichotī) and transporated to headquarters (Anya)

Amber		*Anya*	*Chatsu*		*Anya*	*Bahatri*		*Anya*
Year	*Bichotī*	*(anī)*	*Year*	*Bichotī*	*(anī)*	*Year*	*Bichotī*	*(anī)*
1677	94	6	1711	97	3	1710	96	4
1688	56	44	1712	91	9	1713	97	3
1689	50	50	1714	86	14	1726	59	41
1690	58	42	1724	71	29	1727	99	1
1712	60	40	1726	40	60	1728	92	8
1715	76	24	1730	70	30	1729	89	11
1716	75	25	1731	66	34	1730	98	2
1718	88	12	1733	54	46	1732	77	23
1720	88	12	1735	80	20	1736	83	17
1723	34	66	1737	82	18	1737	76	24
1724	15	85	1738	39	61	1738	67	33
1725	80	20	1740	91	9	1742	100	—
1728	75	25	1741	90	10	1743	100	—
1729	100	—	1742	99	1	1744	100	—
1733	73	27	1744	69	31	1745	100	—
1734	88	12	1745	68	32	1748	97	3
1737	68	32	1748	75	25	1749	100	—
1744	59	41	1749	85	15	1750	49	51

At the village level grain was sold by the government to mainly four kinds of persons, viz.. the *baniā*; *mahājan*, *telī* and *ra'īyat*. The *jama'bandi* documents are very useful in providing us the details of the sale of grain to individuals (see Table II) Clearly, the *mahājans* are quite prominent as revenue-grain purchasers.

But we also see that the disposal of *jinsī* (state share) could often be through the peasant (*ra'īyat*) himself. It is not, however, clear whether by this is meant that the peasant

3. This Table has been prepared from the *arhsattas* of *parganas* Amber, Chatsu and Bahatri.

Table II

Sale of grain in different villages along with the name of purchaser

Village	*Pargana*	*Year*	*Crop*	*Name of the purchaser*
1	2	3	4	5
Dogra	Mauzabad	1654	*Bajra*	*Ra'īyat*
			Jowar	*Awar*
			Til	*Awār*
			Mung	*Sah* Bhan
			Moth	*Awār*
			Singhara	*Baniā, qaṣba* Mauzabad
			Wheat	*Tehvīldār* Than Singh
			Barley	*Awār*
			Ghan	*Mahājan of the qaṣba*
			Sarso	*Ra'īyat*
Dungri Kalan	Kotla	1664	*Bajra*	*Ra'īyat, ahīr*
			Jowar	*Panch* Ransi Ka
			Jowar	*Panch* Ransi Ka
			Ghan	*Vann (bania)*
			Til	*Tēlī* and *baniā*
			Moth	*Baniā* and *panch*
			Urd	*Panch* Ransi
			Barley	sold to Govil Harde
			Gram	Chuhar etc. *Panch* Karamsi Ka
Dara Shikohpur	Liwali	1664	*Kharīf* & *Rabī'* crops	*Ra'īyat*
—	Dausa	1664	*til, moth*	Narwad etc. Baksa
			Kodu	*baniā*
			Barley & Wheat	Bharmal Megh Raj
Biramwas	Nainwa	1665	The whole crop of *kharīf*	*Mahājan* of the *qaṣba* and *ra'īyati*
			Rabī crops	Keso etc. of *qaṣba*
Surtania	Fagui (Phagi)	1698	*Bajra*	*Mahājan*
			Jower	*Ra'īyat*, Sah Behari of village Chakwara

1	2	3	4	5
Sherpur	Malarna	1664	*Jowar* & *Moth*	Devi Das of Sherpur
			Bajra	Gujarmal Hari Das, *mahājan* of Sherpur, Har Ram *mahajan*
			Til	Deepa *baniā* of *qaṣba ra'īyat*, Pitamber of *qaṣba* and Todar etc *telī*
			Barley (*kankūt*)	*Ra'īyat*
			Sarso, *til* (*kankūt*)	Hardya Ram Shankar, *mahājan* of *qaṣba* Malarna
			Moth	Kalu *mahājan*
Akora	Kotla	1663	Barley	Ram Das and Dalwa, *baniā* of the same village
			Til, *Kapas Bajra* & *Jowar*	Mohan Das of Idar Lal Das, Harji and Jagannath of the village Faridabad, *pargana* Kotla
Amarpuro	Mundawer	1665	*Chola*, *Jowar Mung*, *til*,	Bawla etc. *mahājan* of the *qaṣba* Mundawar.
			Moth	Dhana *mahājan* of *qaṣba* Banawar
Ghansri	Au	1694	*Bajra Til*	*Panch*, *mahājan telī* of the *qaṣba*

himself was asked to pay the cash equivalent of revenue due from him, or whether the superior right holders like *patel*, *patwārī* and other *khud-kāsht* peasants were required to collect the grain due and pay purchase money to the State. Sometimes indeed, our evidence is more specific ; and the name of the purchasing *ra'īyat* is given along with the date, month, place and year of the prevailing prices.[4]

4. *Jama'bandī* documents of different *parganas* for different years during the second half of the 17th century, R.S.A. Bikaner. The details of individual villages are given in these documents, eg. in *arhsatta* Chatsu *VS* 1768/1711. The grain was brought to the *khāta* (grain heap). From there, it was sold through the *paṭel* (*ra'īyat*) of the village.

The peasant, too, at times acted as trader carting his produce even from the *ganj* (local market) to the *qaṣba* (town).[5] We may assume that such peasants were not ordinary peasants but must have been *khud-kāsht* peasants, *zamīndārs*, or *ijārādārs*.[6] Some of them were totally exempted from *rāhdārī* (tax on routes from *ganj* to *qaṣba*) and exactions of the *chaukīs* (checkposts) while some others were charged at varying rates. The following details are illustrative of the levies imposed on the carriage of grain from *ganj* to *qaṣba* (*ḥaqīqat ganj ka nāj sadamandsu*).[7]

(a) When an *ijārādār* brought his grain from *ganj* to *qaṣba* he would be exempted from *rāhdārī*; but in case he carried it outside the *pargana* he would be charged.

(b) When a *zamīndār* brought grain from *ganj* to *qaṣba* for sale, he would be charged. From 1730 onwards, this tax had been taken in *ijāra* by Sanghi Ram Singh s/o Jadu Ram, who agreed to pay in half of the amount of *rāhdārī*.

(c) When a *zamīndār* had sold the grain belonging to the state at the *ganj* and had cleared all state dues and then wanted to sell his own grain, he should be charged customary toll tax by the *muṣharrif chauki* (officer in charge of the checkpost).

(d) The *chaudhurī*, *qānūngo*, *muftī* (Muslim jurist), *chobdār* (clubman) and *kalāl* (distiller) were exempted from *māpā rāhdārī*.

Our evidence throws a good deal of light on the functioning of the village market and inland trade.[8] A study of price data suggests rather more stable conditions of the market than one would have supposed. An examination of the range of prices prevailing in the different villages of the same *pargana* at

5. *Yāddāshtī 'amal dastūr mapa rāhdārī pargana* Khohri *VS* 1781/1724; *arhsatta pargana* Naraina *VS* 1787/1730.
6. *Ibid.*
7. *Yāddāshtī 'amal dastūr māpā rāhdārī pargana* Khohri *VS* 1781/1724.
8. *'Amal dastūrs, jama'bandīs arhsattas* and *nirkh-bāzars* of different *parganas*.

the same time indicate that the range of variation in prices as between different villages was not large (Chapter III, Table III).[9] One can assume that at least within the limits of *parganas* the grain market was practically unified. Further, the overall picture of prices which emerges from the study of price data in other *parganas* indicates a marked uniformity in the relative prices of all food-grains barring cash crops.[10] The range of fluctuation of relative prices among the *parganas* over a decade is not very significant. This leads one to the supposition that the unified market extended from the limits of single *parganas* to whole groups of *parganas* constituting practically a sub-region.

We are able from our records (*nirkh bāzār*) to study fluctuations in copper and gold prices in relation to silver.[11] The overall picture of a fairly extensive unified market is supported by the evidence we have of the very close proximity of gold and copper values in terms of silver in the different parganas (see Appendices B-1 and B-2 on pages 290-96).

The rural market was a prominent feature of the local economy in Eastern Rajasthan. Even in the small villages rice, gram, flour (wheat and barley), *ghee*, *maka*, sugar, vegetables and other commodities could be obtained for conveyance to other villages and then to the towns.[12]

A study of *'amal-dastūr rāhdārī* and *'āmal-dastūr chabutra kotwālī* documents, however, gives the impression that a very large number of peasants were not able to reach the open market at all. They, therefore, had no choice but to sell their

9. See my article, 'Prices and Rural Commerce In 17th Century Eastern Rajasthan', *PIHC*, Kurukshetra, 1982, pp. 271-82.
10. *Ibid.*
11. See Appendix 'B' of the book (pp. 290-296) for the silver value of gold and copper for different *parganas* over a number of years. Also see S. Nurul Hasan and Satya Prakash Gupta, 'Prices of Food-grain . . .' Table 6.
12. *Jama'bandī* documents of village Akora, *parganas* Kotla *VS* 1720/1663, 1721/1664 ; village Ghansri, *pargana*, Au 1751/1694, *'Amal dastūrs* and *arhsattas* of different *parganas*.

goods to the village merchants.[13] These local merchants bought commodities from the villages and sold them either in the same village where they had purchased them or in the villages around.[14] The purchasers might be urban merchants or their agents, or even the village moneylender. Our evidence permits us to identify some of the local and urban merchants and to attempt a rough classification of them.[15]

1. *Bichhaitī* or *bachhāyat* : The outside trader who did not live in the village but established temporary residence in the village or who was a pedlar.
2. *Sahthani* : Grain dealers on a modest scale who were involved in trade at the level of the village or the *qaṣba.*
3. *Veopari* or *byopārī* : Big merchants, who operated at every level.
4. *Baniā* and *mahājan* : Trader and moneylender at the level of the village and the town.
5. *Banjāra* : A class of bulk traders involved in transport trade in all variety of commodities. They were not necessarily grain merchants.
6. *Sahardar* (*Shahardār*):[16] A class of merchants who operated in the towns. They most probably purchased goods from the *bichhaitī*.

13. *Dastūr-ul 'Amal* and *'amal dastūr pargana* Phagi 1748/1691 ; *'amal dastūr kotwalī chowtra qaṣba* Phagi *VS* 1770/1713, *'amal dastūr bhomi, qaṣba* Phagi *VS.* 1741/1684 ; *Yaddashtī dastūr-ul 'amal* and *'amal-dastūr ḥāṣil rahdari pargana* Phagi *VS* 1772/1715, *Dastūr Rāhdārī pargana* Malpura (undated), *'amal dastūr māpā pargana* Khohri *VS* 1781/1724 ; *'amal dastūr ghascharai VS* 1769/1712 ; *'amal dastūr pargana* Toda Bhim. *Dastūr-ul 'Amal mahsūl Manāī pargana* Didwana, *sarkār* Nagaur *ṣūbah* Ajmer; *'amal dastūr pargana* Barauda. See my 'New Evidence On Agrarian And Rural Taxation. In Eastern Rajasthan (17th-18th Centuries)', *PIHC*, Aligarh 1975, pp. 229-242. Cf. Dilbagh Singh, 'Nature And Incidence Of Taxes Levied On Inland Trade in Eastern Rajasthan during the 17th and 18th Centuries', *PIHC*, Bhubaneswar, 1977, pp. 311-326.
14. *Ibid.*
15. *Ibid.*
16. *Dastūr-ul 'Amal māl-o-jihāt sāir jihāt pargana* Didwana (undated).

Besides these we have also in *'amal dastūr* references to *dalāls* or brokers involved in local trade.

All those who were directly involved in local trade had to pay a large number of cesses depending upon the commodity. These cesses were levied at various *chaukis* (toll stations, check posts) on way to the market and at the trade centres, though many traders were exempted from these levies.

The following comments may be offered on some of these cesses.

1. *Dastūr-chaudhurī* : Official perquisite paid by the traders to the *chaudhurī* of the *pargana* headquarters-1 paisa on food-grains and 2 paisa on commodities like raw cotton, *gur*, sugar, oil-seed, tobacco, etc.
2. *Rāhdārī* : A toll tax levied by the state. For the details of rates see Tables IV, V, VI and VII.
3. *Māpā* : An important cess levied by the State upon the traders. The exemption from this tax granted to certain categories of persons is also mentioned in the documents (see Tables VI and VII). The collection of *māpā-rāhdārī* was generally taken on *ijārā* by the *mahājan*, *bohra* or rich peasants.
4. *Korī bhomī kī* : Customary tax levied by a *bhomiā* upon the traders passing through his territories.
5. *Batvālī* : The *batval* was an officer concerned with *chabūtra kotwālī* in the *qaṣba*. His main duty was to take station on the road to collect transit dues.
6. *Kotwālī*. The *kotwāl* being an officer in charge of the *qaṣba* was entitled to this perquisite.

   Chabutra-Kotwālī : The main office of the *kotwāl* under whom were different *chaukīs* (check posts), where the trader had to pay the cess on the basis of the nature of commodities (See Tables IV, V, VI and VII for the taxation on different items under different check posts).
7. *Baithak* : A cess levied upon local traders and artisans who brought commodities to sell in the *hāt* (temporary markets in different villages on different week days). A cess known as *hātwārā* also seems to have been of the same nature.

8. ***Bharoti*** : A cess upon the trader while loading his carts. Rates depended on the quality of the commodity being loaded.
9. ***Pasaru*** or ***Nikasu*** : A cess on entry and exit of the commodities. *Nikasu* was the duty levied on goods passing out of a town or a *pargana* ; while *pasaru* signifies the taxes on imports.
10. ***Singoti*** : Charged from those who dealt in cattle, in the weekly markets and fairs.
11. ***Bahetawanī*** : A cess on outgoing and incoming commodities in the *pargana*.
12. ***Khoonchi*** : A cess upon the exchange of commodities between two dealers. It may also be treated as a commission.
13. ***Kayālī*** : Charged from the grain merchant at the time of weighing his grain. *Kayal* means weighman. The rates charged from the *ra'īyat* and merchant were different. (See Chapter VI). At every *chaukī chabūtrā-kotwālī* and markets of the town the government appointed the weighers.
14. ***Tehbāzārī*** : Levied upon those sellers who put their temporary shops in the *bazars, hāts* and fairs ; paid either in money or kind as the rent of the ground on which their shops or stalls were placed.

'Amal-dastūrs of different *parganas* and *qaṣbas* provide us with a very interesting list of the rates of taxation on individual commodities and the categories of persons who either were charged at a concessional rate or were exempted.

There were instructions to the officials to act according to the *paṭṭa* in case the *byopari* carried one. The *chaudhurī, qānūngo, purohit*, Dul Ram, *Nagarseth* and some traders involved in the trade in ghee were exempted from *rāhdārī*. In general *rāhdārī* was to be charged according to the nature of the carriage. In case the cart was half loaded then the charges were half of the *rāhdārī*.[17]

17. *Nakhkhas* was a market for horses and cattle.

Table IV

INCIDENCE OF TAXATION

'Amal-dastūr Chotra (*Chabūtra*) *Kotwālī*
Qasba Phagi (1684 A.D.)

(In *takās* (T) and *dāms*[18] (D) per rupee)

	Seller		*Buyer*	
	Takā	*Dām*	*Takā*	*Dām*
	(1)	(2)	(3)	(4)
Commodities purchased and sold by the local merchants of the same village	—	6	—	12½
Commodities sold by *bichhāit* (outside merchant) in the village	—	31	—	0
Sale of *mahua* by the				
(i) Village merchant to another village merchant of the same village	1	—	—	—
(ii) *bichhāit* to the village merchant or vice versa	1	—	1	—
(iii) *bichhāit* to another *bichhāit*	2	—	2	—
Jinsī sold by the merchants (*khūnch*) to the *bichhāitī*	—	6	—	25
Commodities sold by a *bichhāit* to another *bichhāit* in a village	—	31	—	25
Tobacco				
(i) Sold and purchased by local merchants	—	—	1	—
(ii) Sold by *bichhāit* to local dealer or vice versa	1	—	1	—
(iii) Sold by the *bichhāit* to other *bichhāit*	2	—	—	—
Khūnchīs were exempted from the levy of *kotwālī* on *mahua* and tobacco				

18. One takā carried 50 *dāms* in Eastern Rajasthan. This *dām* is totally different from Akbari *dām*, which was equivalent to one-fortieth of a rupee.

	(1)	(2)	(3)	(4)
On the sale of iron, wood, *sunn*, raw cotton, *moonj* sold to *bichhāit* by the village merchant	—	25	—	—
Iron sold by a local merchant to another local merchant	—	12½	—	—
Khūnchī exempted from *kotwālī*				
On the sale of cattle sheep and goat (*chaillie*)	1	—	—	—
Sale of grains by the *banjārās* coming from East (*purbī binjārā*)	1	12½	—	—
Khūnchī exempted				
On sale of land within the village	3	—	—	—
From the distiller (*kalāl*) in the village per distillary	7	—	—	—
On the sale of *kharbūza* '*khandesari*' etc. per cart by the *bichhāit* in the village	1	—	—	—
On the sale of grass etc. per cart by the *bichhāit* in the village				
(i) *chokhori*	1	12½	—	—
(ii) *beri*	—	31	—	—
On marriage (*bagdam*, *byah gharecha*)	—	—	—	—
The *Mahajan*, Brahman, Rajput, Kayath, Charan, Bhat, Bojig, Jogi, *Dhobī*, Musalman, were exempted from marriage tax (*bagdam*)	—	—	—	—
Sale of animals by local merchant to another local merchant	1	—	—	—
Sold by *bichhāit* while purchased by the local merchant	1	—	—	—
Sc 'd by *bichhāit* to other *bichhāit*	2	—	—	—
Sale of calico and silk cloth by the *bichhāit* in the village	—	6	per *thān* (roll)	
On the oil pressed per head (each *ghān*)	—	25	—	—
Sale of animal skin by *khatik*				
(i) Skin sold and purchased by the local merchant of the same village	—	12½	—	—

	(1)	(2)	(3)	(4)
(ii) Sold by the *bichhāit* to the local merchant of the village or vice versa	—	25	—	25
On each cart of lead and zinc, the impost was not charged ; but the collection had begun during the preceding two years	Rates are not given			
On exchange of the commodities in *hatwāra* by the *bichhāit*, on both *byopārī* and *dalāl*	—	12½	—	—
Purchased by *bichhāiti* from the *ra'īyat*	—	25	—	—
Levy of *kotwāli* on *jinsī* is exempted up to Rs. 4/-. But to be charged after every Rs. 20/- @ *takā*	1	—	—	—
Tobacco, *Mahua*, Jute were exempted from the levy.				

Table V

Dastūr Rāhdārī Pargana Malpura

(per cart per *chaukī*)

A.	*Rasiva*	*Takā*	*Dām*
	Carriage carrying salt *kirāna*, ghee, oil etc.	—	6
(i)	*Bichhāit*	—	12½
(ii)	*Qaṣba* kā	—	9
	cloth, *jasad* (zinc), *sīsa* (lead)	—	25
	Blue colour (indigo) brought by the *kāfila* of Gujrati and Akbarabad (Agra) *khānd* and sugar	1	—
(i)	*Qaṣba* kā	—	12½
(ii)	*Bichhāitī*	—	19
	Til, black and white	—	9
B.	*Sirwa* containing Betel leaves		
(i)	(big *sirva*)	—	25
(ii)	(small *sirva*)	--	12½
	Cloth	—	12½

C.	*Batwalī* levied per *chaukī* in the village. *Qaṣba* exempted from *batwālī* Raw cotton per cart per *chaukī* grain, salt, *til* and cloth per cart per *chauki* Rasiva/20	—	25
D.	*Banjārā* on 100 pack oxen *kirāna*, cloth, *til* etc.	—	3
E.	Camel, buffalo carrying *kirāna*, and raw cotton	—	6

Further details are as follows (on *chauki 1*) :

A.	*Item per cart*	*Rate*	
	Zinc, lead and cloth	Annas	7
	Raw cotton, ghee, oil, iron,	*Takā*	1
	Cotton	,,	2
	Raw sugar (*khand*), sugar and betel of Nagarbel	Annas	6
	Grain, salt, cloth	*Takā*	1
	Kirana, chilli	,,	6
	(i) *Bichhaiti*	Annas	6
	(ii) of *qaṣba*	*Takā*	6
	On indigo, silken cloth brought by the *kafila* (Caravan) from Gujarat and Akbarabad etc.	Rs. 1 Annas	4
	Til, black and white	*Takā*	2
B.	Commodities carried by camel and buffalo on *chauki* 1.		
	Cloth and betel (nagarbeli) *kirana*, raw cotton, *ghee*, oil, *khand*	*Takā*	2
	(i) of *Qaṣba*	*Dām*	37½
	(ii) *Bachhait*	*Takā*	1
	Grain, and salt	*Dām*	25
	Khand and sugar	*Takā*	1
		Dām	25
	Indigo, silken cloth, etc., brought by *kafila* Gujarat and Akbarabad	*Takā*	3
C.	Commodities carried by bullocks		
	Raw cotton, kirana, ghee, oil, *gur*, *khand*, iron, paper, tobacco		
	(i) *Bichhaiti*	*Dām*	25
	(ii) *Qaṣba* kā	*Dām*	19

Cloth, lead, zinc	*Takā*	1
Betel (*nagarbeli*)	*Takā* *Dām*	1 6
(i) *Asli*	*Taka*	1
(ii) *Kayali*	*Dām*	6
Til white	*Dām*	19
Grain and salt from *banjara* and *byopari*	*Dām*	12½
Indigo from the *kafila* from Gujarat & Akbarabad	*Taka* *Dām*	1 25
Khand and sugar (i) of *Qaṣba*	*Dām*	25
(ii) *Bachhait*	*Dām*	37½
Banjara who brought *jinsi*, salt (sambhar) was charged on the numbers of bullocks	*Taka* *Dām*	4 25
The *banjara* of the *qaṣba* was charged at a lower rate.		

Table VI

Rate of *Māpā* per Rupee (of Value) *Pargana* Khohri 1724 A.D.

	Taka	*Dām*
Commodities sold and purchased by *gaventi* (village resident)	—	12½
Commodities sold by *bichhait*	—	19
Commodities brought by *bichhait* from the *qaṣba*	—	12½
Horses etc. sold by *nakhkhas*[1] or corruptly *nakas* (horse dealer)	—	25
Cloth brought by the *bajaj*	1	—
Printing of calico by Syo Singh	Exempted	

Table VII

Pargana Chatsu 1712 A.D.[19]

A.	*Bichhaiti*		(*Taka* and *Dām* per rupee) per carriage, *chauki* 4			
	Items		*Rate*			
	Raw cotton, *ghee*		*Rs.*	*As.*	*Taka*	*Dam*
	Raw cotton, *ghee*	*'aṣli*	1	—	2	—
		kharch	—	4	—	—
		parav	—	—	6	25
	Indigo	*'aṣli*	5	—	—	—
		kharch	—	8	—	—
		parav	—	—	6	25
	Sugar, *khand*, and *kirana*	*'aṣli*	4	8	—	—
		kharch	—	8	—	—
		parav	—	—	6	25
	Gur, khand, til	*aṣli*	—	—	20	2
		parav	—	—	6	25
B.	*Sahthani* (*chauki* 1)					
	Gur, khand, multi, til		—	—	5	—
	Grain		—	—	2	—
	Raw cotton, and ghee		—	8	—	—
	Khand, sugar and *kirana*		1	—	—	—
C.	*Banjara*					
	Rice, *til, naj, gur, khand*	*'aṣli*	2	4	—	—
		kharch	—	4	—	—
	Khand, sugar, *kirana*	*'aṣli*	4	8	—	—
	(*chauki* 4)	*kharch*	—	8	—	—
	Bullock (empty)	*'aṣli*	1	2	—	—
		kharch	—	4	—	—
	Bullock (salt)	*'aṣli*	1	12	—	—
		kharch	—	4	—	—

The following persons were listed in *pargana* Chatsu (1712 AD) as being either totally exempt from the above levy or allowed to pay at the rate of one half.

19. *Dastūr-ul'Amal ghascharai pargana* Chatsu *VS* 1769/1712.

Total Exemption

1. *Chaudhurī qānūngo* including their *birādarī*
2. Jiwan Das
3. Khiwsi *chaukāit*
4. Hari Bhagat *chaukāit* Bhandari
5. Sanghi Sadanand and Mahanand
6. Bhartha and Bhikhari Das
7. Dudo Brahman of Raja
8. Roopchand
9. Ramji Sobhani

One-half Exemption

1. Sundar Sobha *Ṣarrāf*
2. Raja Ram *Ṣarrāf*
3. *Patwārī* of *qaṣba* Patan and Godha inclusive *birādarī*
4. Harji Khasar
5. Sakta *pā'i*
6. Har Ram of *garh*
7. Narain Lali
8. Tara Chand Soni
9. Isar Ajmeri
10. Kusal of Isar

In the welter of exemptions and the complexity of commodities carried, the state was anxious that it should not lose through fraud. Those privileged traders who were exempted from the levies, might pass off goods of the unprivileged as their own. When reports of this reached the state from *pargana* Lalsot, *dīwān* Srichand issued a *parwana* requiring the concerned officials to charge the tax in entirety (*dar-o-bast*) even from those who were enjoying the privilege of exemption of the *māpā-rāhdārī*.[20]

Keeping in view the nature of market, i.e. a unified market of Eastern Rajasthan, one can easily imagine the importance of the role of these local traders, notably the *mahājans* or *panch mahājan* purchasing the *jinsī* directly from the peasant and then selling it in the market. Moreover, the disposal of state *jinsī*, (collected as land revenue) further enhanced their position and enabled them to make profits at both ends.

20. *Parwana* dated *Mangsīr Vadi* 5, *VS* 1770/1713 of *dīwan* Srichand quoted in *arhsatta mujmil pargana* Lalsot *VS* 1770/1713.

Unfortunately, our records do not help us much in determining the nature of long distance trade but for the evidence that indigo and silk cloth were imported from Gujarat and Agra in the territories of Eastern Rajasthan. We can infer that some quantity of the Bayana indigo must have been sold to merchants engaged in long distance trade.[21]

From the information brought together in this chapter it can be seen that there was fairly brisk agrarian commerce in Eastern Rajasthan during our period, although we cannot speak even remotely of the absolute quantities involved. The revenue system clearly was the one most important factor fuelling the trade ; and, in turn, it was itself sustained by the availability of merchant capital and expertise. The mercantile classes were, therefore, an important element of the agrarian economy of the region. Indeed, even parts of the upper elements of the agrarian population, the *zamīndārs* and the *ra'īyat*, found it worthwhile to engage in trade to the extent of buying up the revenue-grain or going to the main *qaṣba* market to sell their produce. The village was hardly an isolated unit in the sense that it is often supposed to have been by authors concerned with the conditions under British rule.

21. See Pelsaert, *Jahangir's India*, pp. 10-18, for conditions of indigo trade early in the 17th century.

CHAPTER V

The Structure of Land Rights and the Agrarian Society

1. *Village Society and Stratification*

The smallest unit of land revenue administration was the *mauẓa'* (village). According to details recorded in our documents,[1] a village comprised *bastī* (area under habitation), *rāha* (tracks, roads), *magro* (rocky land), *pahār* (hills, if any), *nāla* and *sīr* (land under special revenue arrangements), etc., and, finally, the cultivable land (*lāyaq-zarā'itī*).

The records further classify the villages into *'aṣlī* (lit. original) and *dākhilī* (lit. entered). The *'aṣlī* was the older village or one already lying within the limits of the *pargana*, while the *dākhilī* was one which had either been transferred from any adjoining *pargana* or had been newly founded by detaching lands from an *'aṣlī* village. The limits of a village were thus subject to much change over a period of time.

By another official classification found occasionally in documents, villages are grouped into *ra'īyatī* and *ta'alluqa*. We encounter this classification in *pargana* Malarna,[2] where

1. *Muwāzāna dahsāla pargana* Niwai from *VS* 1789/1732 to *VS* 1798/1741, *Taqsīm-i dahsāla pargana* Dausa from *VS* 1787/1730 to 1798/1741 ; *taqsīm dahsāla tappa* Jhak, *pargana* Phagi from *VS* 1782/1725. *Taqsīm pandarahsala* (fifteen years) from *VS* 1706/ 1649 to 1720/1663, *pargana* Antela Bhabhra, *sarkar* Alwar, *Suba* Akbarabad.
See especially the *muwāzana* documents of various *parganas* for different years listed in the Bibliography.
2. *Arhsattas pargana* Malarna *VS* 1783/1726, 1784/1727, 1786/1729 and 1778/1730.

however, the number of *ta'alluqa* villages is almost insignificant in comparison to the *ra'īyatī*.[3]

The distinction between the *ra'īyatī* and *ta'alluqa* villages is that the revenues of *ra'īyatī* villages were collected by agents (revenue collector) of the Mahārāja, while the revenues of the *ta'alluqa* villages were retained by the *ta'alluqadār*, who was essentially a *zamīndār* and who paid a lump sum in lieu of the revenue to the state.[4] Sometimes a group of villages formed for administrative purposes was also known as *taluka* (Pers. *ta'alluqa*).[5]

In the *ra'īyatī* villages there used to be a special class (comprising *paṭels* and others) within the agricultural community, which enjoyed certain superior or *zamīndārī* rights, while the ordinary cultivators (including even 'Rajputs' other than Shekhavats) did not have similar rights.[6]

The *dastūrs* or tax regulations of the *ra'īyatī* and *ta'alluqa* villages were separately issued to *chaudhurīs* and *qānūngos*.[7] The revenue of the *ra'īyatī* villages was paid on the basis of detailed assessment. The revenue collectors had to abide by the state regulations for assessment and collection of land revenue.[8] In contrast the revenue of the *ta'alluqadār's* villages was paid in a lump sum, and no detailed assessment was made.[9]

Beyond the existence of intermediaries in the *ta'alluqa* villages, we can detect within every village, whether *ra'īyatī* or not, a certain degree of stratification. It is not possible, with the evidence at hand, to reconstruct fully the structure of the village society, with the rights and obligations of all the strata

3. *Ibid.* Out of 148 villages only 4 were *ta'alluqa* villages.
4. *Ibid.* Also see *chiṭṭhī kartik sudi VS* 1780/1723. The latter document suggests that the *zamīndārs*, who sometimes contracted with the state to collect the revenue of a specified territory and in that capacity acted as intermediaries between the state and the peasants, were known as *ta'alluqdārs*.
5. *Chiṭṭhīs VS* 1722/1665, *VS* 1802/1745 ; *Parwāna Miti Pos Vadi* 3 *VS* 1782/1725.
6. *Arhsatta pargana* Malarna *VS* 1787/1730 ; *Chiṭṭhīs VS* 1722/1665, 1802/1745 ; *Parwāna Miti Pos Vadi* 3, *VS* 1782/1725.
7. *Chiṭṭhī Miti Pos Sudi* 1 *VS* 1774/1717; *Chiṭṭhī Miti Asarh Sudi* 14, *VS* 1799/1742.
8. *Arhsatta pargana* Malarna *op. cit.*
9. *Ibid.* Also see *arhsatta* Malarna *VS* 1784/1727.

or classes clearly defined. What is however, possible is to identify certain categories of people who appear in our records, and to use these descriptions for throwing light on the social differentiation in the villages.

Official documents introduce us, first, to a simple division of cultivators into (a) *khud-kāsht* (Pers. *khwud kāsht*) literally, meaning 'self-cultivating', but meaning actually those (especially superior right-holders) who cultivated themselves or had it cultivated by hired labour ; and (b) to *ra'īyatī kāsht* (peasants cultivating their fields) and *pā'ikāsht* (non-resident cultivators). Our documents also refer to a term *gaonveti* (resident cultivator), which includes both the *khud-kāsht* and the *ra'iyatī-kāsht* peasants.[10]

The term *khud-kāsht* conveyed a certain element of occupancy right. Its possessors were also sometimes termed *kisān* or *asāmī ra'iyat*.[11] They often had their own *kothi* (granary), well and *ghar* (house) in the village where they lived ; they cultivated with their own ploughs (*gharoo-hala*).[12] Their duty was to sustain cultivation and to meet the state demand for revenue.

The *khud-kāsht* cultivator's holdings were sometimes assessed at a concessional rate.[13] For example, in *pargana* Amber, the state demand on their lands amounted to only one-fourth of the total yield, while other ordinary cultivators (*ra'iyatī-kāsht*) paid the revenue varying from one-fourth to one-half.[14]

Another category, that of the *pā'is* (*pahis*) comprised those peasants, who cultivated lands in villages other than their own.

10. *Arhsatta pargana* Amber years *VS* 1772/1715, 1773/1716, 1775/1718.
11. *Chiṭṭhī Miti Pos Sudi* 1, *VS* 1722/1665.
12. *Chiṭṭhīs Chait Sudi* 9 years *VS* 1722/1665, 1798/1741 ; 1799/1742. Also see *kagad dīwān VS* 1798/1731.
13. In our documents *gharoo-hala* is generally identified with *khud kāsht* signifying a person, who cultivates with his own ploughs and bullocks and possesses some occupancy rights.
14. *Arhsattas pargana* Amber years *VS* 1772/1715, 1773/1716 and 1775/ In 1718. *VS* 1773/1716 in *pargana* Amber, *khud kāsht* cultivator pay the revenue at the rate of one-fourth, while the *ra'īyatī-kāsht pays it at the rate of* 40% in *kharīf* and 30% in *rabī'*. Also see *sanad Baisakh Vadi* 4, *VS* 1823/1776.

Their rights differed from region to region.[15] They were sometimes encouraged by the state to come and settle in new villages and, in such cases, were provided with ploughs, bullocks, seed, manure and money either directly by the state or through a *bohra* (money-lender).[16] On condition of paying the revenue they could retain their fields as long as they wished. In times of distress or famines special instructions were issued to *pargana* authorities to extend patronage and assistance to the *pā'is*. In normal times an attempt was made to induce the *pā'is* to take up cultivation of waste lands or to populate new villages or such villages as had been deserted due to natural calamities.[17]

The *pā'is* do not seem to have been very numerous, and many villages had none. In *pargana* Niwai, out of 41 villages, only 3 had *pā'is*.[18]

In order to induce the *pā'is* to cultivate new land, *pargana* officials were vested with a number of powers.[19] They could grant *paṭṭas* of *vat-vih* (sharing of crop) at a concessional rate, usually of one-fourth of produce, to *pā'is* coming from other *parganas* or adjacent areas.[20] Such grants of *paṭṭas* were generally made for *parat* or waste lands.[21]

15. Satish Chandra, 'Some Aspects of Indian Village Society in Northern India during the 18th Century : The position, and Role of the *khud-kāsht* and *pāhī-kashta*., *IHR*, Vol. I, 1975. In the Rajasthani records examined by me, the term is spelt *pāi'-kāsht* or *pāy'-kāsht* cases where *pāī* can also be read pāhī by a flourish of the scribe are very, very rare. It is, therefore, rather to be presumed that the term arises out of the clerks' Persian, *pāi* carrying the sense of bottom, and *pāi'-kasht* seems those who came at the bottom of the list of the *asāmīs*. See Wilson, pp. 387-88 (S.V. *pahi-kāsht* and *pāī kāsht*).
16. *Chiṭṭhī Sawan Vadi* 10, *VS* 1723/1666 ; 1782/1725; *Bhadva Vadi* 1, *VS* 1762/1705.
17. *Chiṭṭhi VS* 1750/1693 ; *arhsattas pargana* Malarna *VS* 1756/1699; *VS* 1777/1720.
18. *Yāddāshti Bail Hal pargana* Niwai *VS* 1767/1710. Also see Satish Chandra, *op. cit.* The table (prepared by Dilbagh Singh) presents information relating to an exceptional village, where their-*pā'is* number was the highest. 19% of the total.
19. *Chiṭṭhīs VS* 1722/1665 ; 1762/1705.
20. *Chiṭṭhī VS* 1782/1725.
21. *Ibid.*

Another class of cultivators who seem to have enjoyed a lower status than the *khud-kāsht* peasants but probably formed the majority of cultivators were known as *pāltīs*, *pālta* or *pālatiya*. The term *pālti* has been used in our documents frequently but in an apparently different context. Its literal meaning is obscure. Consequently, it is necessary to bring out its sense from the way it has been used in our documents.

Here *pāltī* generally denotes a cultivator belonging to a low caste and enjoying an inferior status in village society because of this.[22] But the documents of *ḥaqīqat hal bail pargana* Chatsu for the year 1666, while enumerating the names of the *'aṣlī asāmīs* (original cultivators) list among the *iẓafa* (additonal) *asāmīs*, the *pālatiyās* including among them Brahmans, Rajputs, Banias, and Jāts along with people of low castes. In the villages of Kottchawack, Dhabbar and Thikara in *pargana* Chatsu, for example, the *pālatyās* do not appear among the *'aṣlī* cultivators. Therefore, *pālatiyās* would seem to mean new agriculturists within the village. We may thus treat the *pāltīs* as migrating cultivators or temporary cultivators.

Another document *dastūr-ul 'amal pargana* Antela-Bhabhra, *sarkār* Alwar, *ṣūba* Akbarabad, speaks of *pāltīs* as the revenue paying class. The *pāltīs* are treated here as part of the *ra'iyatī* (*ra'itī* in Rajasthan) paying 50% of *māl* (land revenue as produce) on *polach* (best or most fertile) land. They thus paid at higher rates than such superior right-holders as Rajputs (specially Shekhawats), *mahājans*, *chaudhurīs*, *qānūngos*, *paṭels* etc.) who were granted concessions.

A variety of *chiṭṭhīs* provide us with miscellaneous information, throwing a good deal of light on the actual position of the *pāltīs*. A *chiṭṭhī* of 1692 from the *pargana* officials to Mahārāja Bishan Singh reports that "the land in *mauẓa* 'Syohar, *ṭappa ḥavelī* (*pargana* headquarters), is lying fallow (*parat*). Therefore, I have invited the *pāltī* of *mauẓa* 'Halsurya, *pargana* Chatsu etc., and founded a new *mauẓ'a'* (village) naming it Biriam ka Nagar."

22. Cf. *Chiṭṭhīs VS* 1722/1665 and 1749/1692.
Also see Satish Chandra and Dilbagh Singh, 'Structure and Stratification of Village Society in Eastern Rajasthan', *PIHC*, Muzaffarpur, 1972, pp. 196-203.

It may be noted that here we have a case of colonization or the founding of a new village. The cultivators with whom it is to be settled are styled *pāltīs*; they were apparently not *pāltīs* in their own villages, but would become such when induced to move to newly settled village, presumably under the inducement of some concessions. The concessions extended to them by the state were the grant of the *paṭṭa vatvih* (concessional *paṭṭa*), and provision of ploughs, manure, seeds, etc.

There is some very interesting information which some of these papers offer under the heading *iẓāfa* in each village.[23] The *iẓāfa* signifies additional peasants including *pāltīs* being settled together with their cattle. The statement made under this shows that migration generally took place from the neighbouring *qaṣbas* and *parganas*; some migrants had previously been "servants" (*naukar*) elsewhere. But some were previously wage labours (*majūrs*), who had now obtained bullocks and were able to set up as peasants. Quite obviously, land was available ; it was the possession of cattle which was the crucial pre-requisite for starting cultivation. So far as we can see, once the *pāltī* had settled, his rights over the land he occupied were the same as those of the older peasants.

The function of the *pāltī* was no less significant in the days of famine and scarcity. During such periods, they proved to be an asset for the state. A *pargana* official (*'āmil*), Kesodas wrote to *dīwān* Kalyandas in 1665, "The *ra'iyatīs* (probably used for *gharoohalas* or the persons owning their own ploughs *khūd-kāshta* and *gaonveti* or resident peasants) have gone to some other places in famine years. Therefore, I have instructed all the *chaudhurīs* and *qānūngos*, *patels*, and *paṭwārīs*, *pāltīs* and *ijārādārs* of the *parganas* Bahatri, Piragpur Jalalpur and Bharkol, who have gone to Malwa, Burhanpur, Kalibhit and *Purab* and Pilibhit towards Agra and Shri Muthra (Mathura) to console the *pāltīs* and to attract and give them inducements for *paṭṭa baṭā'ī* (concessional sharing by *paṭṭa*),

23. See *Yāddāshtī haqīqatī hal bail* of different *parganas* ; *Raqbo pargana* Mauzabad *VS* 1698/1641.
See my art., 'Agriculture and Agrarian Society, Mid 17th Century Information From Records of *Pargana* Mauzabad In Rajasthan', *PIHC*, Bodh-Gaya, 1981, pp. 370-76.

bullocks and seed. I (Keso Das) have also granted concessional *paṭṭas* to the *paṭels* of two or four village, who have gone to bring the *pāltīs* back."

Thus we see that the best opportunity the *pāltīs* to improve their lot and raise their status lay in days of scarcity, when, if fortunate enough to survive, they could by the use of the resources of the state or *bohras* (on behalf of the state), emerge as independent cultivators.

The above discussion suggests that the *pāltī* or *pālatiyā* formed a class of migrant cultivators who seem to have generally enjoyed a lower status as compared with the *khud-kāsht* peasants, but who could sometimes be treated at par with *ra'iyatī* or *khudkāshtī* peasant. The reason for styling them *pāltī* was possibly because the revenue authorities extended their patronage to them, giving them various concessions. In other words, the key word in *pāltī* may be *pal*(*a*), to patronise. In the process of expansion of cultivation, settlement of a new village or resettlement of one deserted, the *pāltīs* might have received the ultimate benefit of permanent occupancy rights identical with those of *khud-kāsht* cultivators.

Up-till now there has been no evidence that the *pāltīs* could sell or give their holdings on farm (*ijāra*) to others. This perhaps formed an important difference between them and the *khud-kāsht* peasants.

Besides the three designated groups of peasants, there were those who had no right over lands or crops. They were labourers, being sometimes called *gunwars*. They generally belonged to non-peasant castes, such as barber, *telī* (oil-presser), *mālī* (gardener), etc. They could be compelled to render forced labour (*begār*) for agriculture by the upper village strata (*bhomīa*, *paṭel*, *jāgīrdār*, etc.).[24] The precise nature of their

24. *Chiṭṭhī VS* 1798/1741 of Sita Ram etc. *qaṣba* Mauzabad to *faujdār* and '*āmils pargana* Mauzabad.
Chiṭṭhī *Mah Vadi* 3, *VS* 1801/1744; *Chiṭṭhī Jeth* Sudi 15, *VS* 1802/1715 *Chiṭṭhī VS* 1799/1742, *Chiṭṭhī VS* 1798/1741.
Also see Satish Chandra and Dilbagh Singh 'Structure and Stratification'. While stating the composition of the peasants in *pargana* Wazirpur *VS* 1783/1726, particularly of 4 villages (pp. 198-199), op.

obligations and rights, if any, is unfortunately not clear owing to the paucity of data.

The stratification of village society was also reflected in the system of taxation. A section among the peasants who paid land revenue at concessional rates enjoyed certain privileges.[25] The land revenue was imposed at much heavier rates upon the *ra'īyatī* and *pāltī* than upon the upper strata.

A perusal of Tables I, II and III[26] shows the incidence of land revenue demand under *baṭā'ī-jinsī* (crop sharing) upon different classes of the village revenue-payers. They show that generally the land revenue-demand was fixed at a half of the produce in respect of all crops except wheat and *bajra*, where it was three-fifths. For favoured revenue payers, however, such as the Brahmans, *mahājans*, *Paṭels*, and *bhomias*, the land-revenue demand ranged from one-fourth to two-fifths. The *pā'is* were either totally exempted from taxation or were asked to pay it at much lower rate, probably in view of their being invited to till virgin land.

cit., they are put in three categories. The third category belongs to service classes such as potters, carpenters, gardeners, ironsmiths, gold smiths, tanners, oil-pressers, washermen, barbers, etc. The percentage of the number of their households out of the total was 12%. For *begār* and *baithi begār* see Harbans Mukhia, 'Illegal Extortions from Artisans and Menials in 18th Century Eastern Rajasthan *IESHR*, Vol. XIV, No. 2 (April-June 1977), pp. 231-45.

25. See my art., 'The System of Rural Taxation in Eastern Rajasthan (c. 1665-1750)', *PIHC*, Muzaffarpur, 1972. Also see Dilbagh Singh, 'Caste and the Structure of Village Society in Eastern Rajasthan During the 18th Century', *IHR*, January 1976, No. 2. He makes the interesting observation that the *gaonvetis* were subdivided on the basis of caste into two categories : the *riyayatis* ('concession-holders' from *r'āīyat*, concession) and the *ra'iyatis*. In accordance with the customary practices of the *pargana*, the *riyayatis* were the privileged class while the *ra'īyati* constituted the unprivileged section. The term *ria'yati*, however, does not appear in the documents that I have seen. Even the word *ri'āyat* (lit. concession) is not met in the revenue records.

26. These Tables are prepared from the *dastūr-ul 'amal pargana* Phagi, Amarsar, Thek, Udai and Bhairana. Also see *dastūr-ul-'amal* and *'amal-dastūr pargana* Phagi *VS* 1748/1691 and *arhsattas* of different *parganas*, where the rates of state land revenue demand upon different classes of peasants have been given.

Table I

Crop-sharing (*baṭā'ī*). *pargana* Amarsar, *sarkār* Nagaur, *ṣūba* Ajmer, 1726*

					Rajputs				Sikar Khetri *zamīndārs* (*bhomias*) & *chaudhuris*,	FARAH (Jihāt)	
Crops	*State*	*Pāltī* & *Ra'īyat*	*Brahmans*	*Pā'i*	*Shekhawat*	*Others*	*Patel* etc. *Kamīn*	*Mahājan*	*qānūngos*	*Seri*	*nirani*
Kharīf	50%	50%	—	—	33%	33%	—	33%	—	*mahājans* exempted	Rajputs exempted
Sali (Rice)	40%	60%	33%	—	—	—	40%	33%	25%	*Paṭel* is exempted from *jihāt*	
Singhara	40%	60%	exempted from *jihāt*					exempted from *jihāt*	exempted from *jihāt*		
Madhwa	40%	60%	—	—	—	—	—			seri & (*seer*)/3/- per *man* from *ra'īyat*	*dastūr chaudhuri qanūngo ser*/1/per *man*.
Rab'ī											
Gram	50%	50%	—	—	—	—	40% exempted) (from *jihāt*	—	—	—	—
Barley	50%	50%	—	—	—	—	—	—	—	—	—
Wheat	40%	60%	33%	—	—	—	40%	40%	25%	—	—
Bajra	40% instead of 50%	60%	—	—	—	—	—	—	—	—	—

**Dastūr-ul 'Amal pargana* Amarsar *VS* 1783/1726.

Table II

Dastūr, *Pargana* Udai, *Sarkār* & *Ṣūba* Akbarabad*

State Share

Crop Sharing (*baṭāī*), *Pargana* Udai

	Raiyatī	*Qānūngo*	*Mahājan*	*Kheldār* etc. *Paṭel*	*Rajputs*	*Patel Pa'i*	Brahman, Bhat *Kisan of qaṣba Udai*	*Pa'i*	*Kamīn Kisān*	(*Naqdī*)
Kharīf :										
(i) *Polach* land	50%	25%	40%	33%	33% 25% (kotri Rajputs)	40%	50%	40%	33%	3% Exemption to *Qānūngo* Brahman, Bhat, *mahājan* *Paṭel* and *Kamīn*
(ii) *Banjar*	40%	—	—	—	(kotri Rajputs)	—	—	—	—	
Rabī'										
(i) *Polach*	50%	25%	40%	33%	30% 25% (kotri Rajputs)	40%	50%	40%	33%	3% Exemption to *qānūngo* Brahman, Bhat, *mahājan* *Paṭel* and *Kamīn*.
(ii) *Banjar*	40%	—	—	—	--	—	—	—	—	

**Dastūr-ul 'Amal pargana* Udai *VS* 1772/1715.

Table III

Dastūr,[27] *Pargana* Bhairana, *Sarkār* & *Ṣūba* Ajmer, 1723

State Share under Jinsī (*Crop-Sharing*)

Asāmī	*State Share*	*Jihāt* (*seri*)
Paṭel	40%	exempted
Kamīn	40%	exempted
Ra'īyati	50%	not exempted
Chaudhurī, *Qānūngo* & *Kayath*	25%	exempted
Musalman, *Nagauri Deswal*	50%	Some are exempted and some
and some of them	40%	have to pay according to the *khasra* of the *paṭwarī*.
Rajputs		
(i) *Gaonveti* (*kotridār*)	25%	exempted
(ii) Other Rajputs residing in the same village	40%	exempted
Some of them	33%	exempted
Bohra Har Dutt	25%	exempted
Banias	one-eleventh	some are exempted
Some of them	40%	some have to pay according to the *Khasra* of *paṭwarī*.

Over and above the *māl* (land revenue) the peasant had to pay *jihāt* (which included transport charges and other perquisites : see Chapter V). The other superior revenue-payers were again exempted from this burden.

Thus a section among the revenue-payers belonged to the privileged class, who not only paid land revenue at concessional rates but were exempted from the *jihāt* tax also known as *bandhatī*, approximating to 15% of the amount realized in *naqdī* and also *seri* which ranged from 12.5% (5 *sers* per *man*) and 25% (10 *sers* per *man*). The rate of *jihāt* varied from place to place. The other ordinary cultivators had to bear the maximum burden of taxation.

It seems that the ordinary peasant, whether *ra'īyati* or *pāltī*, was under considerable financial pressure. On the other hand,

27. *Dastūr-ul 'Amal, pargana* Bhairana, *sarkār*, *ṣūba* Ajmer *VS* 1780/1723.

the *mahājans, bhomias, chaudhurīs* and *qānūngos* (*pargana* officials), *paṭels* (village official) and a class of Rajputs (mainly the Shekhawats), belonged to the privileged class within the village. The tendency of the official agrarian policy was clearly designed to perpetuate or even increase the stratification within the village, by making the rich richer and the poor poorer through a system of discriminatory taxation.

The extent of the economic stratification that had been reached can be judged from data on the size of holdings, number of crops sown by individual peasants and the cattle possessed by the different revenue-payers within the village. Such information comes to us from certain documents known as *Khasra*,[28] though in some cases the period is a little later than one would have wished.

The earliest *Khasra* (*Khasro*) traced relates to *batā'ī* (crop sharing) of village Tuta, *pargana* Tonk, for the *rabī'* 'harvest of 1711, the crop shared being barley. It gives the name of the *asāmīs* (individual revenue payers), the total produce of the crop of each, the state share in the produce (ranging between 33 to 40 per cent). (See Table IV). The size of tax paid by different individuals is an indication of the range of differentiation. Three peasants respectively obtained yields of 129, 126 and 111 *mans* (the latter two being *paṭels*). As many as 19 *asāmī* out of 39, however, obtained yields of less than 10 *mans* each. Yet the smallest producers (8 in all; producing less than 10 *mans* each) were made to pay tax at 40% while all the others paid at 33.3% of the produce.

There is an interesting entry under the designation of *panchāiti*, credited with 270 *man*. (taxed at 33.3%) and 9 *mans* 5 *sers* (taxed at 40%). This could have been the land nominally sown by cultivators (outsiders?) in the name of the village community' or of its dominant 'council' (*panchāit*). Unluckily in spite of the significance of the term the nature of the '*panchāitī*' land remains obscure.

A *Khasra* of *qaṣba* Soabdaspur, *pargana* Sawai Jaipur, 1791, *Kharīf* harvest covers the area of 9 *paṭṭīs* ('strips). This was

28. For *Khasra* documents, see my article, '*Khasra* Documents In Rajasthan' in *Medieval India—A Miscellany*, Vol. IV, 1976, pp. 168-176. In Kota the same details were recorded in the *Khasra*, *tazkara* and *Khāta* documents, with minor differences.

Table IV

Khasro Batā'ī, Mauẓ'a Tuta, *Pargana* Tonk,
Rabī', 1711 A.D.

BARLEY

State (*dīwān*) share
(One-third in *mans*.)

Asāmī (Cultivator)	*Jumlo* (total produce in *mans*)	*Ra'tī* (cultivator)	*Dīwān* (State)
Hardya	39.30.0	26.20.0	13.10.0
Kalo	60. 0.0	40. 0.0	20. 0.0
Tor	58.35.0	39.10.0	19.25.0
Kusalo	129.30.0	86.20.0	43.10 0
Kasto	69.30.0	46.20.0	23.10.0
Dharamsi	3. 0.0	2. 0.0	1. 0.0
Badri	64.20.0	43. 0.0	13. 0.0
Govind	39. 0.0	26. 0.0	13. 0.0
	24.30.0	16.20.0	8.10.0
Sirsa	66.30.0	44.20.0	22.10.0
Dharamsi	55.20.0	37. 0.0	18.20.0
Parso	15. 0.0	10. 0.0	5.0 .0
Deva	71.10.0	47.20.0	23.30.0
Udai Singh	15. 0.0	10. 0.0	5. 0.0
Syama Barber (*naee*)	18. 0.0	12. 0.0	6. 0.0
Sama Jat	3.15.0	2.10.0	1. 5 0
Hardo	24. 0.0	16. 0.0	8. 0.0
Har Ram (F.)	84. 0.0	56. 0.0	28. 0.0
Bhagwan	1.31.0	1.10.0	0.25.0
Hiro	0.30.0	0.20.0	0.10.0
Hajila	0. 6.0	0. 4.0	0. 2.0
G. Tej Singh	33. 0.0	22. 0.0	11. 0.0
Sig Ram	72. 0.0	48. 0.0	24. 0 0
Pirag *patel*	126. 0.0	84. 0.0	42. 0.0
Narain *patel*	111. 0.0	74. 0.0	37. 0.0
Kaledha *dhobī*	9.30.0	6.20.0	3.10.0
Nanu *dhobī*	6. 0.0	4. 0.0	2. 0.0
Dudo	1.20.0	1. 0.0	0.20.0
Tejo	1. 5.0	0.30.0	0.15.0
Devo	0. 9.0	0. 6.0	0. 3.0
Radho	0.12.0	0. 8.0	0. 4.0
Kanho	0.12.0	0. 8.0	0. 4.0
	1206. 9.0	804. 6.0	402. 3.0
Panchāitī	270. 0.0	180. 0.0	90. 0.0
Total	1476. 9.0	984. 0.0	492. 0.0

Barley *bāta pachaduvō*
(Two-fifth), in *man*

Asāmī	*Jumlo* (Total)	*Ra'īti* (Peasant)	*Dīwān* (State)
Kisto	1.20.0	0.36.0	0.24.0
Ram Chand	10. 0.0	6. 0.0	4. 0.0
Dharmo	0.10.0	0. 6.0	0. 4.0
Hiro *patel*	5. 0.0	3. 0.0	2. 0.0
Lave *gujar*	2.10.0	1.14.0	0.36.0
Puro	12.20.0	7.20.0	5. 0.0
Devo	0. 2.8	0. 1.8	0. 1.0
Hiro	0.15.0	0. 9.0	0. 6.0
	31.37.8	19. 6.8	12.31.0
Pānchāitī	9. 5.0	5.19.0	3.26.0
(Total) :	41.28	24.25.8	26.17.0
Asāmī (Head)	*Jumlo*	*Ra'ītī*	*Dīwān*
Vato tisro (one-third)	1476.40	984. 6.0	482. 3.0
Vato Pachaduvo (two-fifth)	41.3.0	24.26.0	16.17.0
Terij			508.20.0

originally a large joint or coparcenary 'holding', comprising many villages. Many of the *asāmīs*, especially *paṭels* and also some other superior right-holders had their holdings in more than one *paṭṭī*. There were some small and big *paṭṭīs* containing 4 to 55 *asāmīs* respectively. Here again, we see a considerable degree of differentiation prevailing.

As a sample, let us take the *paṭṭī* of Sukhpura. Out of 42 cultivators a single *asāmī* cultivated 9 crops of the *Kharīf* season; 4 *asāmīs*, 7 crops; and 7 *asāmīs*, 6 crops each. All these 12 cultivators may be supposed and belong to the top stratum cultivating between 6 and 9 crops which included cotton, *til* and maize. Six of these 12 cultivators were *patels*. Of the rest, 11 cultivators cultivated 3 or 4 crops each, comprising the middle stratum. The remainder, 19 in all, sowed one or two crops only, and thus stood at the bottom.

Details of other *paṭṭīs*, however, suggest that the cultivators cultivating more than 3 crops generally belonged to the top stratum. These were mostly *paṭels* or *mahājans* or belonged to some superior castes or clans (especially Brahmans and Rajputs). There were some large holdings in the *paṭṭīs* by some lower caste people as well. For example, in the *paṭṭī* of Rampura, out of 19 *asāmīs* belonging to different professions (such as barber, *cotton-carder*, potter, boatsman, weaver, etc.), 4 *asāmīs* cultivated 3 to 5 crops each. One potter had raised 5 crops. In the *paṭṭī* of Maharajpur, a barber cultivated 5 crops, which also included maize and cotton. This position of some of these low caste cultivators may possibly have been exceptional since we are here concerned with a *qaṣba* or township.

As for caste-wise distribution of peasants in *paṭṭī* Manigpur, 75% of the revenue-payers recorded were Brahman, while in another *paṭṭī* Rampur 76% of cultivators belonged to the lower castes.

There is, however, one important reservation to be made here. We have assumed that persons cultivating one or two crops were 'peasants'. It seems that many of these were only 'part-time' peasants. Some of them might even have been sharecroppers (*sanjhādārs*) and tenants, while the other could have been temporary cultivators. The reasons for this assumption are : firstly, the area under cultivation is not recorded against their names; secondly, the land revenue paid by them in kind is very small, suggesting a very minute holding; and thirdly, even the one or two crops cultivated by them are of a very inferior kind.

For another village, though from Kota territory and of a later date, we have data on the size of holdings. In village Piplod, *pargana* Tajpur, 1808, *Kharīf* harvest, there are recorded 122 *asāmīs*. Of these, 4 held between 10 to 20 *bīghas*. Of the top, four villagers one was a *paṭel*, one a *mudro Mārwari* (*baniā*), and two *telis*. 11 cultivators cultivated from 5 to 10 *bīghas* each; and 24 cultivators only from 1 to 5 *bīghas* each. The remainder—as many as 83 villagers—cultivated less than one *bīgha* each. From the same *Khasra* of village Piplod it appears that about half the *asāmīs* cultivated both *kharīf* and *rabī*' harvest.

Finally, we have a large number of documents giving the number of bullocks owned by individual peasants thus supplying us important data about the heads of cattle (bullocks) per peasant.[29] We take two sample villages, viz. Jhak and Pachal in *pargana* Mauzabad in 1641. (See Table V). The richer peasants here owned up to 9 bullocks each and the poor, one or two bullocks only.[30] It will be seen that certain castes or categories, e.g. *patels*, were more liberally supplied with cattle.[31] Clearly, traditional power and caste lay behind much of the stratification.

Table V

Peasants and Bullocks (Summary Statement from Official Survey)

(i) *Mauẓa'* Jhak

Asāmīs 116

Bullocks 320

(A) *Caste-wise break-up*

Asāmī		*Cattle*
Patel	6	28
Paṭwarī	1	9
Jat	7	16
Ahir	14	38
Na'i	3	5
Gujar	4	8
Brahman	6	12
Kumbhar	2	4
Teli	2	2
Chipro	2	10
Kharwal	5	12
Miscellaneous (caste not given)	62	185

29. *Raqbo pargana* Mauzabad *VS* 1698/1641 ; *hāzrī asāmī wa bail pargana* Mauzabad *VS* 1723/1666 ; *Yāddāshti haqīqatī hal bail zubani paṭel paṭwārī pargana* Chatsu *VS* 1723/1666.
30. If we look into the number of bullocks possessed by *paṭels* in other villages, the number goes up to 14 bullocks owned by one *paṭel* on the average. (Also see Chapter II)
31. *Ibid.*

(B) *Number of peasants by ownership of bullocks*

Bullocks	*Asāmīs* (Peasants)
9	1
8	1
7	3
6	4
5	8
4	8
3	15
2	54
1	20

(ii) *Mauẓa*ʿ Pachal

Asāmīs—30

Bullocks—96

(A) *Caste-wise break-up*

		Cattle
Paṭel	2	12
Jāt	12	30
Brahman	5	15
Khati	1	2
Miscellaneous	10	35

Asāmīs	(B) *Bullocks (per asāmī)*
1	8
1	5
7	4
9	3
12	2

For *qaṣba* Chatsu (1666), we again have a caste-wise break-up of the *asāmīs* as well. The *asāmīs* here were 200 in number and the number of bullocks owned by them was 422 (2.11 each). The break up was as follows :

Qaṣba Chatsu, 1666

Caste	*No. of asāmīs*	*No. of bullocks*	*Bullocks per head*
Mali	70 (4 *paṭels*)	123	1.76
Gujar	8	18	2.25
Teli	29	53	1.83
Brahman	53 (3 *paṭels* and) (2 *chaudhurīs*)	119	2.14
Miscellaneous	40	109	2.75
Total	200	422	2.11

Our sources thus suggest that at the village level the *paṭels* were big peasants, who must have organized *k͟hud-kāsht* cultivation with hired labour, while at the opposite end were the small peasants engaged in cultivation who could hardly afford to own ploughs and cattle. If one were to use the language of our documents, we might be tempted to designate the top peasants as *k͟hud-kāsht* and the remainder as *paltīs*. But since *k͟hud-kāsht* was very probably a semi-legal inheritable right, possessed also by some small cultivators, such direct identification of the legal categories with strata demarcated on the basis of economic differentiation may only be of broad validity, subject in detail to many exceptions.

CHAPTER VI

Superior Rural Classes

I. *The Bhomīas*

We investigate in this chapter the structure of rural rights above that of the peasants. This involves, principally, a scrutiny of the rights of the *bhomīas* (*bhūmiās*) who would seem to be the equivalent of Mughal *zamīndārs* in Rajasthan and were to be found in all parts of Rajasthan.[1]

Tod treated the word *bhomīa* to mean "allodial proprietors" perhaps keeping in view the "allodial" vassals of the European feudal system.[2] Subsequent research into the rights and privileges of *bhomīas* in Rajasthan has left little doubt that the *bhomīas* resembled the Mughal *zamīndārs* in all essential particulars, though the *zamīndārs*' rights and privileges might differ from region to region.[3] In contemporary documents too the *bhomi*

1. In Western Rajasthan, the *bhomīa* class appeared under the Rathors from the time of Rao Chunda, when he subdued some independent principality whereupon the leaders of the dislodged ruling families were recognised as *bhomīas* of their respective lands. Matrimonial relations were established by the Rathore rulers with the dislodged Rajput families. The villages held by the *bhomīas* were known as *bhomī chara ra gaon*. G. D. Sharma *Rajput Polity* : *A Study of Politics and Administration of the State of Marwar*, 1638-1749, pp. 4 and 20n.
2. James Tod, *Annals and Antiquities of Rajasthan* (Delhi, 1971), i, p. 33.
3. Cf. Irfan Habib *Agrarian System*, p. 139 ; G. D. Sharma *Rajput Polity*, op. cit.; B. L. Bhadani, 'The Allodial Proprietors' (?)—The *bhūmias* of Marwar, '*IHR*, July 1979-Jan. 1980, Vol. VI, Nos. 1-2, pp. 141-153 ; S. P. Gupta and Shireen Moosvi, '*Bhomī* in the Territories of Amber' *PIHC*, 1970, pp. 353-60.

(*bhūmī* or *bhom*) is treated as synonymous with *zamīndārī*.[4]

The grant of *bhom* was considered to be a mark of special favour by the state. So cherished was the tenure that the chiefs and sub-assignees of chiefs used to be eager to obtain it even in villages wholly dependent on their authority.[5] The right was more secure and inheritable than other grants or *jāgīr* assignments.

Our documents suggest that the majority of the *bhomīas* in the territory of Amber belonged to Rajput clans such as the Narukas, Kalyanots (Kilanot), Rajawats, Khangarots, Chauhans, Panchanots etc. But there were also, occasionally, *bhomīas* belonging to other castes, viz., Jāts, Minas, Jadams and Panwars, among others. Different Rajput clans dominated particular *parganas*. Thus the Narukas held *bhomī* rights mainly in Malpura, Phagi and Bahatri ; Chauhans in Udai, Hindaun and Lalsot ; Shaikhwats in Khandela, Khetri, Jhunjhunu, Fatehpur, and Uniara ; Solankis in Tonk and Niwai ; and so on.[6]

It appears that from the point of view of their origin there were two types of *bhomīas*. First, those who had emerged on the basis of their *thikāna* or conquest made by their ancestors and whose *jamīndārī* rights prevailed in one or more villages on the basis of their caste or clan affiliations. The rulers of Amber were themselves such *bhomīas*. Second, there was a class of *bhomīas* created by the rulers of Amber, who were granted *bhom* land (village or group of villages) for some special reasons. Attempts were made by the latter group of *bhomīas* to also assume the position of *jāgīrdārs* on behalf of the Amber rulers.

4. *Arhsatta zamīndarī pargana* Mauzabad and Bhairana *VS* 1752/1695. Also see *arhsatta bābat zamīndāri bhom, pargana* Balheta, Bahatri, etc.
5. James Tod, *Annals and Antiquities*. Vol. I. p. 138.
6. This information derives from the '*arẓdāsht* and *chiṭṭhīs* of different *parganas* for different years. Cf. R. P. Rana 'Agrarian Revolts in Northern India during the late 17th and early 18th Century', *IESHR*, XVIII (3 & 4), pp. 291-326 and Annexure 1 to his article for the areas of *zamīndārī* of different caste and clans.

The *bhomīas* often maintained *gaṛhīs* (forts) in their *bhom* villages.[7] And the state reserved its rights to grant permission *garhībhom* (raising a fortress) to a new person in case there was no *zamīndār* (*bhomīa*) in the village.[8]

On how the *bhomīa* actually obtained his income there is tantalisingly little information. In a *chiṭṭhī* of 1666, it is stated that the *bhomīas* levied their claim on a share of the produce separately from that of the State. Before 1631, the *bhomīa's* share out of the *ra'iyatī* share under *jinsī* was *sers* 2 per *man*, or 5% of the portion of produce left with the peasant after the payment of the tax. But after 1631, it is claimed, it was reduced by only a quarter *ser* per *man* or by 0.55% of the produce left with the peasant.[9] It may be that this was supplemented by the *bhomīas* by levying other claims and perquisites ; on these, however, unfortunately, there is very little evidence in the documents studied so far.

Possibly, a part of the income of the *bhomīas* also came from lands they either cultivated themselves or got tilled by others under their names. The category of peasants known as *vasidārs* were settled by the *bhomia* in the village, to undertake sometimes the cultivation of his personal lands. In return the *vasidār* usually got a part of the produce or an allotment of a piece of land, on which the assessment was comparatively lower than that of an ordinary cultivator.[10] On these lands the gain of the *bhomīa* came from special tax concessions. Thus in one *paragana*, at least, they had to pay land revenue at the very favourable rate of a quarter of the produce.[11]

The *bhom* villages were assessed by the state usually at a lower rate than other villages, and so a share in the surplus

7. *Chiṭṭhī* of Bijai Singh Khangarot to Maharaja Bishan Singh Ji dated *Miti Asarh Sudi* 13, *VS* 1740/1687 regarding village Mehendwas, *pargana* Tonk.
8. *Sanad, Chait Sudi* 2, *VS* 1644/1787 ; *Chṭṭhī* of Keso Das to *dīwān* Bhikari Das and Fateh Chand.
9. *Chiṭṭhī* of Mukund Das and Keso Das (*āmil* and *amīn*) to *dīwān* Kalyan Das, dated, *Miti, Sawan Sudi* 1, *VS* 1723/1666.
10. *Dastūr-ul 'Amal pargana* Antela *VS* 1794/1734 ; *Jama'bandī pargana* Dausa *VS* 1772/1715.
11. *Dastūr-ul 'Amal pargana* Amarsar, *VS* 1783/1726.

could accrue to the *bhomīas*. There was in addition, a cess known as *bhom*. An *'arẓdāsht* of 1687 suggests that the state was charging *bhom* as per custom at the rate of one *ṭaka* per *bīgha* and one *ser* per *man* in *jinsī* (*maash* and *moth*) from the *ra'iyat*. The *ra'iyat* complained that *bhom* had been levied upon them twice, once by the state, and then Gaj Singh Rajawat (probably the *jāgīrdār*).[12] The amount of *bhom* thus received was generally deposited under the heading *bhomvat*.[13] The rights to *bhom* were usually granted to the sub-assignees by the rulers of Amber. The *bhom* as a cess was an additional right of such *jāgīrdārs* as also held *bhomī* right, In the *arhsatta bhom* or *bhomī*, we get details of the demands under *bhomī* rights. The total of *bhomī* exactions, relatively to the total land revenue realization, was not very significant. They varied from 1% to 3% and 1% to 1.25% of the total revenue in *parganas* Naraina and Malarna respectively.[14] This right to exact *bhom-bāb* could be transferred or mortgaged by the ruler or the *jāgīrdār*, as the case might be.[15] Cases were reported to the Amber rulers about *bhomīas*' resisting the rulers' claims. It seems that the Amber rulers were always interested in enlarging their *zamīndāri* rights and weakening the rights and privileges of the *bhomīas* in order, ultimately, to appropriate the *bhomīas*' rights entirely. Consequently, many *bhomīas* continued to fight, often on a clan basis, against the Raja of Amber.[16] Sometimes

12. *'Arẓdasht Miti Jeth Sudi* 12, *VS* 1748/1687. Our information comes from the *Arhsattas bhomī* of different *parganas*.
13. *Ibid*.
14. S.P. Gupta and Shireen Moosvi, '*Bhomi* in the Territories of Amber', op. cit.
15. *Lakhtang* documents of different *parganas*. The rates of *bhom-bāb* were not fixed. The *bhomīa* could even sometimes be exempted from the payment of *bhom-bāb*.
16. R.P. Rana, 'Agrarian Revolts', *op. cit*.
    See the cases of Pratap Singh Narula of a village Soda in *pargana* Malpura ; the Solankis of Tonk and others, who rebelled for this cause. All the Narulas *bhomīas*, from Mewat to Tonk and Malpura, joined hands against the Raja and his allies to defend their *zamīndāris* or even to create new ones.
    So was the case with the Jāts in Rajasthan and outside : Amar Singh Chauhan, a very powerful *zamīndār* (*bhomīa*) of *ṭappa* Khair and

they appealed at the Mug͟hal Court and brought *parwānas* to assert their rights. Thus in 1666, Mukund Das Keso Das (*pargana* officials) informed *dīwān* Kalyan Das that Ajmeri Khan etc. (*bhomīas*) went to Agra to contact Nawab Zafar Khan and also Shri Maharaj Kumar and brought necessary *parwānas* to confirm their rights.[17]

There was indeed a traditional obligation on the rulers to protect *bhom*-rights. Its possession was considered hereditary and was generally held sacrosanct, being often awarded after the performance of exceptional services. On the other hand, the *bhomīa* was always under an obligation to furnish militia for the service of his overlord.[18]

A large number of *chiṭṭhīs* testify to forcible occupation of the villages, oppression and terror by the *bhomīas*, which enhanced the misery of the peasantry. A *chiṭṭhī* of 1664 by Bhat Shanker (a *pargana* official) to *dīwān* Kalyan Das reports that "the *ra'īyat* abandoned cultivation under the constant threat of *bhomīas* and had gone to Delhi to appeal. This is not

Maharajpur in *pargana* Kol (Aligarh) supported the cause of the Jāt *zamīndārs* and these *zamīndārs* were supported by the peasants as well. This was perhaps the reason why Jai Singh was reluctant in accepting the *jāgīr* of *pargana* Kol in *tank͟hwāh*, and on obtaining it, adhered to the farming system. The farmers (to whom the revenue was farmed out) were naturally the local *zamīndārs*. For a detailed discussion, see my article (unpublished) Revenue-Farmers and *Zamīndārs* in *Pargana* Kol (Aligarh), 1694." Paper presented at the Session of Indian History Congress, Also see *Jama'bandi pargana* Kol (Aligarh) *VS* 1751/1694, Hijri 1101-02 ; '*Arẓdāsht* (*Baisakh Vadi* II, *VS* 1749/1692 ; *Vakils*' Report (Persian) concerning the Mahārāja of Jaipur, Jan 24, 1694.

17. *Chiṭṭhī Miti Sawan Sudi* 1, *VS* 1723/1666.
18. According to Tod, a *bhomīa* could be called for local service in the district (*pargana*) where he had resided for a certain period of time. The *bhomias* constituted a local militia. They held and maintained the important fortress of the district. They paid a quit-rent to the crown (state) and performed local but limited service on the frontier garrison. (*Annals and Antiquities*, *op. cit*., Vol. I, pp. 231, 238, 576-577.) Appendix No. XVIII, p. 243 for the grants of *bhum* land in *mundkati* or compensation for blood. Also see *Vir Vinod* I, pp. 136-137, where it is stated that the duties of *bhomīas* were to keep watch and ward over their particular villages. In case of need, they had to present themselves for any service along with their contingents.

good. On enquiry, I found from the *ra'iyat* (the *paṭel* and the *muqaddam*) that there is no grievance against us (Mahārāja) ; the grievance is against the *bhomīas*. However, the *'amal* (tax-collection regime) is of the Mahārāja."[19] They (*bhomīas*) charged illegal and irregular cesses from the peasants and thus were responsible for desolating villages.[20] Another letter of Girdhar Das and Manrūp (*amīn* and *'āmil*, *pargana* Niwai) informed Mahārāja Bishan Singh that there was famine (*kahat*) in 1694 in *pargana* Niwai. Every attempt was being made to bring the *ra'iyat* back to cultivation, but the *bhomīas* continued to oppress the peasantry in the villages.[21] Besides the peasantry, the *bohras* and traders were also oppressed by the *bhomīas*. Sometimes the *bhomīa* was reported collecting the *ḥaṣil* and *ḥaṣil rāhdārī* (the states' claim) from the traders and *banjāras* (grain merchants).[22] The loot of the traders and harassment of *bohrās* by the *bhomīas* was quite frequent. Consequently, *bohrās* became reluctant to advance loans due to the non-availability of adequate security and uncertainty of returns.[23] A *lakhtang* of 1710, written by Jas Ram, Daf Ram

19. *Chiṭṭhī* dated *Miti Sudi* 1721/1664.
20. *Chiṭṭhī Miti Sudi* 1721/1664 ; *Miti Sawan Sudi* 2, *VS* 1762/1665.
21. *Chiṭṭhī VS* 1753/1696 ; Also see *chiṭṭhī Vadi* 7, 1744/1687. Gopal Das (*'āmil*) informing Maharaja Ram Singh that the *jama'* of *pargana* Amber had been reduced to less than half in the current year (1687) as compared to 1680 and 1681 because of the oppression by the *bhomīas*. They (*bhomīas*) had looted the *ra'īyat* and traders, etc., in the *pargana* and created a lot of trouble. The Jat (*bhomīas*) of *pargana* Baswa supported the cause of Narukas. Other *bhomīas* such as Bargujars, Meos, Chauhans and Jadams supported the Jats. One contingent of the army of Jats came towards *pargana* Baswa ; the other towards Hindaun and was at a distance of three *kos* from *pargana* Udai. People of Baswa and Udai were terrorized to hear this news. The (Mahāraja's) army was stationed in the *pargana* (Amber) ; therefore, the Jats had seized the land elsewhere. The Bargujars (*bhomīas*) also created trouble in the time of Maharaja Man Singh, Bhao Singh, and Madho Singh and fought with them. The *'āmil*, therefore, requested the Maharaja to take proper care of these *parganas* and remove the grievances of the people caused by the *bhomīas* ; only then, would it be possible to extend *khāliṣa* land.
22. *Arhsatta* Malarna *VS* 1723/1616 and 1779/1722.
23. *'Arẓdāsht* and *Chiṭṭhis VS* 1751/1694.

Maf Ram and Raja Ram (all *bohrās*) to the *dīwān* reported that village Gopal Rai Pura in *pargana* Chatsu was founded by their grandfather, named Kheti *bohra*. The whole village was desolated due to the oppression (*khaichal*) of the *bhomīa*. The writers had now been given four plough lands in *in'ām* (roughly 200 *bīghas*) to rehabilitate the said village.[24]

Some efforts were made by the Amber rulers to protect the peasantry and other classes like *bohras* and traders from the oppression of the *bhomīas*.[25] In some cases, the state found it difficult to collect the land revenue from the areas of such turbulent *bhomīas*, where *ra'īyat* also supported the cause of the *bhomīas* against the state perhaps because of caste or clan considerations. This might, in turn, have been one factor which led the Amber rulers to resort to large scale *ijāra* (revenue farming).

II. The Agrarian Mercantile Classes

A different position in the agrarian structure was occupied by the *mahājans* (grain dealers) and *bohrās* (money-lenders). They acted as middlemen between the state and the peasants, and as such exercised considerable control over the rural economy of Eastern Rajasthan.

The *mahājans* (*banīas*), who were primarily grain merchants, were assigned certain obligations in the conduct of trade by the state, e.g. procuring grain from the countryside, and arranging its sale in rural or urban markets, and giving reports

24. *Lakhtang Miti Karti Vadi* 12, *VS* 1767/1710. The land covered under a plough (*hal*) differs from region to region. In Western Rajasthan, for example, 50 *bīghas* of land was deemed to be served by a plough, while in Kota it was 60 *bīghas*. However, evidence is not available for its size in Jaipur for the 17th century.
25. *Chiṭṭhīs* of Harnam to Mahārāja Bishan Singh. Also see a heading of *jami'at-kharch* in different *arhsattas* of different *parganas* for the measures adopted by the state to suppress such *bhomīas*. For example, Rs. 191.50 and *ṭaka* 664 were spent over *jami'at-kharch* to suppress Zorawar Singh and Hari Singh s/o Himmat Singh Rajawat, *bhomīas* who came to *garhi* Pipla and attacked the Haras of village Sunara. They were turned out from that place. *Arhsatta bhomīa, pargana* Chatsu 1691 A.D.

of prices of the agricultural produce.[26] They also played an important role in controlling the prices of various grains sold in the market. The authorities generally displayed some anxiety regarding the regular supply of the grains in the town and its sale at fixed prices even in the days of scarcity. In a letter dated 1705, Kishor Das (*'āmil* informed Mahārāja Jai Singh that "the prices due to scarcity of rains are rising to one and a half times more. I (*'āmıl*) have called upon one *mahājan* not to raise the price. Then he (the *mahājan*) informed me that he had no grain left in store, and that it is *purohit* Shyam Ram, etc., (other *mahājans*) who have hoarded the grain and are selling it at their own prices. I (*'āmil*) instructed them also to sell the grain at the usual rates and not to charge excess prices."[27]

The *mahājans* used to advance loans to peasants to purchase agricultural equipment, to expand cultivation, or to buy seed and means of subsistence in times of scarcity. The state, *ijārādārs*, *jāgīrdārs* and *pāltī* also borrowed money from them for different purposes.[28] They enabled the *ijāra* system to function by standing surety (*māl-i ẓāminī*) for the *ijārādārs*.

There is evidence that the families of the *mahājans* practically monopolised offices in the revenue administration in different *parganas*; and appointments were generally made on the basis of kinship with existing officials.[29]

The *bohrās* (*sāhū* or *sāhūkār*) appear in our documents as moneylenders, who also engaged in issuing and discounting

26. *Chiṭṭhīs VS* 1721/1664 from Akhey Raj to *dīwān* Kalyan Das. *Chiṭṭhī VS* 1722/1665 from Amra Kotwal to diwan Kalyan Das. *Chiṭṭhī Jeth Sudi* 10, *VS* 1749/1692; *Chiṭṭhī Sawan Sudi* 2, *VS* 1762/1705 ; *Chiṭṭhī VS* 1771/1714 from Kaushal Singh, Udai Ram and Anand Ram to *dīwān* Bhikari Das and Feteh Chand. Also see different *tojīs* (bundles) of *nirakh bāzār* (market prices) and *nirakh roznama* (daily prices) which were duly signed by *pānch-mahājans*.
27. *Chiṭṭhī Sawan Sudi* 2, *VS* 1762/1705.
28. See various *chiṭṭhīs* and *lakhtangs* of different *parganas*. Cf. Dilbagh Singh, 'The Role of the *Mahājans* in the Rural Economy in Eastern Rajasthan during the 18th Century', *Social Scientist*, May 22, 1974, pp. 20-31.
29. *Arhsatta pargana* Malpura and Niwai. The office of *'āmil* was retained by *mahājans* in *pargana* Malpura for three generations.

hundīs (bills of exchange). The main function of the *bohra* in a village was to provide *taqāvī* loans (*tagai*) to the peasants on behalf of the state or sometimes on his own.[30] The *bohras* were officially expected to supply seed, ploughs, bullocks and manure to the *ra'iyat* at normal rates.[31] On many occasions the state and other dignitaries (such as *jāgīrdars*, *zamīndārs* and *ijārādārs*) had to depend upon the *bohras* for credit to meet their own financial obligations.[32] The *bohras* was legally entitled to *biyaj-ghiwai* (interest).[33] The rate of *biyaj* and *ghiwai* was Rs. 1.50 and Rs. 2 per cent per month respectively.[34] The names of the *bohras* were recorded, so that some check might be maintained on their operations, especially at the time of crop-failure, when they could greatly oppress the peasants.[35] Thus a *chiṭṭhī* of an *'āmil* addressed to *dīwān* Kalyan Das informs him that "*Purohit* Shyam Das is the leading *bohra* in the *pargana*, who also advanced agricultural loans to the *ra'iyat*. But now that he is refusing to give the loans. I (*'āmil*) may be permitted to use force against him."[36]

In case the *bohras* declined to advance loans to the indigent peasants, the state could order them to do so.[37] There was apparently an understanding between the state and *bohras* to lend money to the state at the rate of ¾% or 1% (*ponotra* or *ekotra*) per month,[38] thereby enabling the state to advance loans to the cultivators. Perhaps, the *bohras* generally preferred this bargain keeping in view the security of the money. A *parwāna*

30. *Chiṭṭhī Sawan Vadi* 10, *VS* 1723/1666; *Chiṭṭhī VS* 1722/1665; *Parwāna Miti Asarh Sudi* 8, *VS* 1783/1726.
31. *Chiṭṭhī VS* 1722/1665 ; *Chiṭṭhī VS* 1725/1668 ; *Chiṭṭhi Miti* Bhadva Vadi 1 *VS* 1762/1705.
32. Cf. Dilbagh Singh 'The Role of the *Mahajans*', op. cit.
33. See different *arhsattas* for the references of *biyajghiwai* ; *Sanad Phalgun Vadi* 11, *VS* 1805/1748, *Chiṭṭhī Jeth* Sudi 2, *VS* 1782/1725.
34. *Parwāna Mah Sudi* 15, *VS* 1787/1730.
35. *Chiṭṭhī Miti Sawan Vadi* 10 *VS* 1723/1766 ; *Chiṭṭhī Miti Asarh Sudi* 8, *VS* 1783/1726.
36. *Ibid.*
37. *Chiṭṭhī* from Pandit Dulta to Shri Girdhar Das Ji *VS* 1723/1666 ; *Chiṭṭhī* from Ghan Raj (*'āmīl*) to Kalyan Das (*dīwān*) *Miti Sawan* Vadi 10, *VS* 1723/1666.
38. *Arhsattas* of different *parganas*.

by Mahārāja Jai Singh Sawai addressed to *dīwān* Vidhyadhar 1725, ordered that "in case the *bohra* declined to advance loan to *ra'īyat* the addressee was to make provisions for it from the state, so that it may be possible for the *ra'īyat* to cultivate land in *pargana* Amber. In no case, should the *ra'īyat* be permitted to cultivate the land of any other *jāgīrdār*. The Mahārāja says that he had already issued *parwānas* in this connection to the *amīns* and *'āmils* to the *pargana* with similar instructions."[39]

The *mahājans* and *bohrās*, thus, formed two distinct, though similar, professional groups. Together, they exercised considerable control over the village economy. They provided capital for the expansion of cultivation, and extorted fairly high rates of return through usury. The heavy demand of revenue in cash and uncertainty of production mainly due to uneven distribution of rains made the peasants very greatly dependent on the moneylenders (*mahājans* and *bohrās*) for their survival. The *mahājans* and *bohrās* must have enlarged their income and wealth at the expense of both the State and the peasantry. But they did not apparently convert their money claims into land rights. Thus, these two mercantile classes rarely appear as *bhomīas*, whose ranks continued to be monopolised by the Rajputs and the superior agrarian castes.

39. *Arhsattas* of different *parganas*, *Parwāna* (Hindi) *VS* 1782/1725.

CHAPTER VII

The System of Rural Taxation

The importance of taxation in the medieval revenue administration has been frequently emphasized. A study of the large number of taxes levied by the rulers of Amber in their territories may be of considerable help in forming a larger conception of the 17th-18th century Indian taxation system. In the *arhsattas* we have a village-wise breakdown of the taxes of the *pargana*, with the rate of demand under each individual item. The detailed figures are, however, not provided for the area assigned in *jāgīr* or granted in *punya-udik* (charitable lands), *in'ām*, etc.; only the lump sum amount of assessed income (*tan*) is given. Since the information is massive, it may be helpful to select a few villages by way of sample and to calculate in each case the net revenue demand and the amounts of the cesses under different heads.

The *arhsattas* divide the *ḥāl-ḥāṣil* (actual revenue collection) of the cultivated land into two categories : *ḥasbul-mufasal* (from both the harvests *kharīf* and *rabi'*) and *ḥasbul waṣūl*.[1] *Ḥasbul-mufaṣal* is further divided into two heads under which the dues of the state were realized : *Muwāfiq-jama'bandi* (according to assessment) and *Siwāi-jama'bandī* (in addition to the assessed revenue).[2]

1. *Ḥasb-ul Waṣūl* is a term used in revenue records to designate items of uncertain value, of which no estimate can be computed, and which are entered only after their actual receipt (Wilson, *Glossary*, p. 201).
   However, in the *arhsattas*, *ḥasbul-waṣūl* indicates the amount realised from those villages which were previously assigned in *jāgīr*, but transferred or resumed to the *khāliṣa* in mid-season.
2. Literally, *Siwāī Jama'bandī* means revenue realised in addition to *jama'bandī*. See below the comments on *jamā'bandī*.

The *Muwāfiq-jama'bandī* is subdivided into *māl-o-jihāt* and *sāi'r-jihāt*. The *mālo-jihat* comprises revenue realized through *zabṭī* (*zabṭi* proper), *qarār-ijāra* (fixed farmed out revenue),[3] *baṭā'ī-jinsī* (crop sharing) and *jihāt* (explained below).

Māl seems to represent the original tax (land revenue) and the *jihāt* the cesses collected to meet the expenses incurred in connection with the assessment and collection of *māl*.

The *jihāt* assessment on *māl* (both in *zabṭi* and *jinsī*) was made up of two heads : (1) *lawāzima*, corrupted into *lawāzimo* or *lozimo*[4] (perquisites mainly in *zabṭi*) and (2) *farah*[5] (in *jinsī*). *Lawāzima* and *'alūfa* were fees or levies paid to different officials or staff of particular establishments, such as the *'āmil*, *'amīn*, *kotwāl*, *chowkāyat-khazāna*, *mushrif chabūtra-kotwāl* or *kotwāli*, *mushrif-bāgh*, *awārdār* (in-charge of granary), *daroga-awār*, *chowkayat*, etc.[6]

Lawāzimo is further sub-divided into *dehnimī zabṭi* (lit. '5 per cent *zabṭi*') at the rate of 5% of the *māl*, and *jaribāna zabṭi bīgha* (sugarcane exempted) at the rate of 1½ *taka* per rupee of *māl*. Sugarcane was exempted from both these levies.[7] The rates add up to the amount given as *jihāt*. The rate of *dehnimi*, i.e. 5% is the same as in the other regions, while the rate of *jaribāna* seems to have been higher in Eastern Rajasthan than in other regions.[8]

3. *Qarār ijāra* denotes that amount of *Khāliṣa* villages or portions of the villages, the revenue of which were farmed out.
4. *Arhsatta* Lalsot 1772/1715. See Chapter VI.
5. *Farah* may be from *fara*, singular of *farūat*. Irfan Habib, *The Agrarian System of Mughal India*, p. 243. Also see Wilson, *Glossary* s.v. *Fara*, 'lit., Above, Over, a term used in the Mohammadan revenue accounts, for the additional revenue derived from new taxes.'
6. *Chiṭṭhī* (undated) of *dīwān* Rai Narain Das and Kirpa Ram to *pargana* officials regarding their appointments as *'āmil* and *amīn* of *pargana* Amarsar and Khatoo. The *dīwān* assures them that the *lawāzima* and *'alūfa* are to be claimed by them, and that they can collect them from a particular *potdār*.
7. *Arhsatta* Malarna *VS* 1787/1730.
8. Noman Ahmad Siddiqui, *Land Revenue Administration Under the Mughals*, Appendix 'C', p. 158. The author has given an example of the assessment account of village Ganeshpur. In this village *jaribāna* and *dehnīmi* both were fixed at the rate of 5%.

Under *jinsī*, the *jihāt* cesses were designated *farah*. From this the *qānūngos*, *paṭels* and other privileged classes were exempted. *Farah* is divided into *divotri* (at the rate of 1 *ser* per *man*.), *ṭappadāri* (at the rate of 1/16 of the *ser* per *man*), *dastūr-qānūngo*, *bhāra* or *farah-bhāra-ganj* and *iẓāfa-ganj*[9] (a charge levied upon the peasants for the transportation of grain to the mart (*ganj*), where it was sold).

Under *ẓabṭī*, different rates per *bīgha* were applicable to individual village. There were changes in the rate of demand made upon individual crops assessed under *ẓabṭī*. Apparently, the rate depended on the quallty of the land,[10] and prices. However, our examination of the rate of revenue demand per *bīgha* for *ẓabṭi* crops has shown remarakble fixity (Chapter II).

We have already seen (Chapter IV) that different rates of taxation were applied to different strata in a village whether under *ẓabṭī* or *jinsī*. Persons enjoying a superior status often paid land revenue at concessional rates. These concessions could vary from *pargana* to *pargana*, but were present in practically all *parganas*. The following, for example, represented the *dastūr* in a select group of *parganas*.

Apart from the village and local officials, there was another privileged class comprising *mahājans*, Rajputs, (Shekhawats) and Brahmans, who were assessed more lightly and liberally.[11] The *pā'is* (non-resident cultivators) were also sometimes granted *paṭṭas* at concessional rates presumably depending upon the availability of cultivable waste land in particular villages.[12]

Under the *jinsī*, the share of the state often varied from one-fourth (*bāt-chautho*), one-third (*bāt-tisro*), two-seventh (*bāt adh-tisro*) and two-fifth (*bāt-pachduvo*), to one-half (*bāt-adho*)

9. *Arhsattas pargana* Lalsot, *VS* 1788/1731; Amber 1721/1664, and Dausa 1776/1719.

10. *Dastūr-ul 'Amal* and *'amal dastūr*, *pargana* Phagi *VS* 1748/1691, *Hijri* year 1099. The rates upon *magro* (hilly) lands were lower than those of *goriwa* (sandy) e.g. the rate of *vani* (cotton) in *goriwa* lands was @ Re. 1/- per *bīgha*, while in *magro* lands it was As. 12/- per *bīgha*.

11. *Dastūr-ul 'amal*, *pargana* Phagi, op. cit. See Chapter IV.

12. *Dastūr-ul 'amal*, *pargana* Amarsar, op. cit.

Table I

Rate of Demand (Under Ẓabṭī)

Pargana	*Year*	*Crop*	*Chaudhurī qānūngo, paṭel Rs. per bīgha*	*Ra'īyatī Rs. per bīgha*
Phagī[13] *sarkār* Ajmer	1715	Sugarcane	1.00	2,50
		cotton	0.25	0.94
		Sunn	—	0.67
		Sakarkandi	0.25	1.37
		Kodo and *Kakri*	0.25	0.75
		Madhwa	0.25	1.75
		Maka Jowar ; *pā'is* according to *paṭṭa*		0.94
Geejgarh[14]	1725	Cotton	1.50	1.75
		Chola	0.50	.73
		Sugarcane	1.75	2.50
		Vegetables	1.25	2.00
Amarsar[15] *sarkār* Nagaur, ṣūba Ajmer	1726	*Jowar*	1.75 (patel, etc.)	
		Maka	1.50 (Rajput)	2.00 (*pālti*)

of the crop.[16] The variations were due sometimes to special categories of revenue payers, different crops or dry/wet land or, finally, different kinds of soil.

It may be seen from the scrutiny of administrative manuals and revenue accounts like *dastūr-ul 'amals* and *arhsattas* of different *parganas* that the share of the state which ranged generally between one-fourth to one-half covers all categories

13. *Dastūr-ul 'amal, pargana* Phagi *VS* 1772/1715.
14. *Dastūr-ul 'amal, pargana* Geejgarh *VS* 1782/1725.
15. *Dastūr-ul 'amal, pargana* Amarsar *VS* 1783/1726.
16. *Arhsattas pargana* Naraina *VS* 1757/1700, 1787/1730. Also see Chapter II.

of cultivators in a village, whether *khud-kāsht* or *pā'i kāsht*, privileged or unprivileged. Broadly speaking, the average state share (*ḥiṣṣa dīwān*) seems to have come to 40% to 45% of the total produce in normal years. In abnormal years such as 1712, 1713, 1715, 1717-18, 1731 and 1737, however, the average does not exceed more than one-third.[17] For example, in 1713, the state share was only 23% in *pargana* Dausa; in *pargana* Lalsot, in 1713 and 1715, it came respectively to 34% and 35%.[18]

An examination of the *dastūr-ul 'amal* of different *parganas* shows that the share of the produce levied on the *ra'īyat* or ordinary peasantry was normally one-half.[19] But it would be unrealistic to generalise and consider it to be a universal or unvaried proportion.

Some of the factors which determined the variations in the magnitude of land revenue demand in crop sharing are brought out in our documents. Different rates of land revenue demand were sometimes prescribed on the basis of irrigation facilities. For example, the crops cultivated in *bara*[20] or *vor* or *vara kyari* (land of the first quality) were charged at the rate of one-half, while the *piwal* or *piyal* (irrigated) lands watered by wells (*dhenkli*) and *tāls* (tanks) were charged at a lower rate. But there were other factors behind the variation as well. A perusal of Table II shows the incidence of land revenue demand in *baṭā'ī* or *bhāolī* in respect of all crops. The Table brings out two striking facts. First, the *ra'iyatīs* and *pāltīs* had to pay one-half as land revenue (from the first quality of land) in respect of all crops except wheat and *bajra*, where it was two-fifths.[21] Second, a part of the produce was appropriated in addition to *māl* under the head of *jihāt*. The *jihāt* was sub-divided into two parts, viz. *dastūr*, *chaudhurī qānūngo* and *farah*.

17. See Chapter II.
18. *Arhsatta* Lalsot *VS* 1772/1715.
19. *Dastūr-ul 'amal pargana*, Phagi, 1691, and *Dastūr-ul 'amal pargana* Amarsar.
20. *Bara* was the land of the first quality, although containing a portion of sand.
21. The rates of land revenue for *bajra* were reduced to two-fifths from one-half and to one-third from two-fifths for those cultivators who had sown it in *rabī'*. See *dastūr-ul 'amal pargana* Phagi.

The rate of perquisites (*dastūr' chaudhurī-qānūngo*) was one *ser* per *man*. Under *farah* two cesses were levied, namely, *serī* or *serīna* ranging from 3 *sers* to 5 *sers* per *man* depending sometimes upon the quality of the crop and *nirani* (transportation charges for carrying grain to the township where it was sold). The charges varied from place to place, according to the kind of crop.

Other revenue payers such as Brahmans, *mahājans* (*banīas*), *paīels* etc. paid land revenue at lower rates under *jinsī*, in the same way as under *ẓabṭī*. Similar concessions again were extended to *pā'is*. Either they were exempted from any tax for a certain period from the time they began cultivation, or the demand was put at quite a low figure from the beginning.

Table II

Division of Produce
Under Crop Sharing

Crop	*Share of State in tax (māl)*	*Palti or Ra'ī-yatīs*	*Seri (Ser per man of States share)*	*FARAH Nirani*	*Perquisites Dastūr Chaudhurī Qānūngos*	*Total tax-burden as % of produce, excluding Nirani*
Kharīf	50%	50%	5		One *ser* per *man*	65%
Sali (Rice)	40%	60%	3 to 4	Transport		47½ to 50%
Singhāra	40%	60%		charges are		47½%
Mandhwā	40%	60%	,,	to be fixed		47½%
Rabī'				according to		
Gram	50%	50%	5	the estima-		62½%
Barley	50%	50×	5	tion under		62½%
Wheat*	40%	60%	3 to 4	the head		47½%
Bajra	40%	60%		*izāfaganj*		47½%

*40% was to be charged from those cultivators who were previously paying at the rate of 50%; those paying 40% in previous years had to pay 33%.

Sāi'r-Jihāt

As explained above. there were besides the *mal-o jihāt*, other sources of tax-income under the heads of *sā'ir-jihāt* and *siwāi-jama'bandī*. A study of the cesses grouped under *sāi'r-jihāt* show that they were imposed as either customary fees or as taxes from the residents of a village, such as transit duties, house tax, market tax, tax on animals, etc. The items of taxation that I have counted from various documents number as many as 88 in all. Some of these taxes are primarily agricultural, while others are non-agricultural. The *sā'ir-jihāt* also included certain impositions, allowances or remunerations payable to certain local officials, grain merchants, etc.

The income derived from the *sā'ir-jihāt* ranged, during different years, from 2% to 25% of the total fiscal revenue in *pargana* Amber, 4% to 15% in Bahatri, nil to 22% in Lalsot, 1.5% to 6% in Malarna, 1% to 3% in Chastu and 2% to 12% in Dausa.[22] Except for Amber (until 1725) the percentage did not normally exceed 10%. In *pargana* Amber too, the range is maintained at 9% to 25% between 1677 and 1724, but is brought down to 2% to 7% from 1725 onwards. Clearly, this reflects the shift of the main urban settlement from Amber to Jaipur. This piece of information accords with the view that the *sā'ir-jihāt* comprised in the main non-agricultural taxes.

The amounts of revenue derived under the various items of *sā'ir-jihāt* are specified in our documents. Sometimes the rate of tax under an individual item is also recorded. The miscellaneous taxes are so numerous that a discussion of the nature

22. The range for Amber is obtained from *arhsattas* for the years, 1677, 1680-90, 1712, 1715, 1716, 1718, 1720, 1723-25, 1728, 1729, 1733, 1734, 1736 and 1737 ; Chatsu for the years 1708-1712, 1714, 1716, 1717, 1721, 1723, 1724, 1726, 1730, 1731, 1733, 1735, 1737, 1738, 1741, 1742, 1744, 1745 and 1749 ; Bahatri for the years 1665, 1684, 1688, 1689, 1696, 1697, 1706, 1708, 1710, 1716-1718, 1720, 1721 and 1723-1725 ; Dausa for the years 1712, 1713, 1715, 1736-1739, 1741-1747, and 1750 ; Malarna for the years 1690, 1711-1716, 1719, 1723, 1726-1730, 1732, 1736-1738, 1743-1745 and 1748 ; and Lalsot for the years 1687, 1713-1715, 1717-1722, 1726, 1731, 1733, 1734, 1737, 1739-1741 and 1745-1748.

and magnitude of each here is likely to put too great a strain on the patience of the reader ; the discussion is therefore relegated to Appendix A to this Chapter. Though we have succeeded in identifying most of the important cesses mentioned under the category of *sā'ir-jihāt*, there are still a few which remain unexplained or uncertain. They are, however, not very significant, in terms of magnitude, and were also realised only rarely. The total revenue realised under such unexplained items ranges from 1% to 5% of the total of *sā'ir-jihāt* realization, and so can be regarded as insignificant.

In order to present some idea of the magnitude of the taxes discussed above, we are giving below the amounts realised under each of the important taxes in the different *parganas*, in terms of percentages (set in ranges) of the total amounts of *sā'ir jihāt*. (See Appendix 'A' for the definition of the taxes).

Pargana Amber : *patwāra* (3% to 42%), *pāni* (11% to 40%), *chailli* (12% to 17%), *māpā* (8% to 84%), *kayāli* (9% to 11%), *jhupri* (9% to 41%), *ghughri shahna*, *jurō* and *tulai*, *ghiwai*, *baqayat* (gardens), *rozīna ṭapdār*, *faroi*, etc.

Pargana Bahatri : *dastūr qānūngo* (6% to 33%), *patwāro* (7% to 38%), *pichotri* (6% to 20%), *sarhi* (10% to 40%), *ghiv* (1% to 5%), *rozīna ṭappadāri* (2% to 9%), *bhent* & *teḥṣīldāri* (1% to 20%), *jhuprī* (40% to 13%), *chailli* (3% to 23%), etc.

Pargana Chatsu : *sarhi* (10% to 60%), *sarhi athotari* (1% to 4%), *patwāra* (10% to 40%), *kayālī* (10% to 51%), *zarībāna* (1% to 15%).

Pargana Dausa : *sarhi* (1% to 18%), *patwāra* (2% to 5%), *tulai* (1% to 8%), *bhāra* (3% to 32%), *ekotri* (1% to 10%), *khari khakhlo* (17% to 46%), *shahna uthai*, (1% to 6%), *jhuprī* (4% to 20%), *chailli* (4% to 12%), *māpā* (1% to 5%).

Pargana Lalsot : The important taxes are *rozīna tappadarī* (10% to 27%), *sugan bhent* (1% to 9%), *bhent khichri* (14% to 40%), *māpa* (2% to 16%), *pāni* (20% to 63%), *chailli* (4% to 26%), *jhuprī* (13% to 42%).

Pargana Malarna : The important taxes are *teḥṣīldārī* (18% to 53%), *rozina tappadāri* (8% to 27%), *norotri* (4% to 36%), *jhupri* (17% to 40 %), *tulai* (5% to 26%).

Besides the above, certain taxes, unimportant in size in these *parganas*, were important elsewhere. Thus, *gaosumārī* (*go-sumāri*) or cattle-tax was an important tax in *pargana* Phagi in 1698.

Siwāī-Jama'bandī

The *siwāī-jama'bandī* appears to represent an addition to the standard revenue demand, whether as an increase in the amount of demand or in the shape of new or additional cesses or imposts levied. More than one hundred items of taxation are recorded in the *arhsattas* of different *parganas* under this head, though the taxes actually entered in individual villages are far fewer in number. These cesses are recorded in villages in the *khāliṣa* as well as *jāgir*. These included numerous additional exactions and perquisites appropriated by officials personally and by the *zamīndārs* and are accordingly excluded from *muwāfiq-jama'bandī*. Broadly, indeed, these taxes may be classified under two categories : (1) perquisites and (2) general taxes.

The important perquisites include the *dastūr-chaudhurī, dastūr-qānūngo, dastūr-patwārī, ḥāṣil chabutra-kotwālī, shahnagi, muqaddamī, rozīna ṭappādāri, peshkash*,[23] *virad sawa* (*daroga chauki*), *parkhai potakhana* (for treasury), *wāqia-nawīs, nānkār*,[24] *chawkidārī, tafāwat* (*nirakh tafāwat mohur*).

The general taxes and fees were : *ghās charai, ṭalabāna*,[25] *farohi, gunjāis, gheraicha, noto*,[26] *agarkhari, khandsal*,[27] *mela*,[28]

23. An 'offering' expected from lower officials, *paṭel, chaudhurī* & other officials, grantees and revenue-payers on occasions, e.g. at the time of foundation of a new village. *Arhsatta pargana* Amber ; also see *sanad* of appointment of *paṭel, Miti Sudi* 7, 1744 A.D. Jaisingh Pura, *ṭappa* khoh, *pargana* Jaipur.
24. The share of the *chaudhurī* and the *qānūngo* out of the revenue approximately 6.82%. *Takhmīna, pargana* Gaonri.
25. *Talab* technically meant claim of salary or allowance from public revenue/treasury. Thus *talabāna* should have been a tax or charge levied on any one drawing a salary or allowance, e.g. on a *tankhwāhdār* (*Arhsatta* Amber).
26. Marriage tax.
27. A tax on the manufacture of salt levied upon the *kharwāl* (manufacturer).
28. A tax on sugar manufacture levied on the merchant.

tehbāzārī,[29] *pānī, bhent lata ẕabṭī,*[30] *malba,*[31] *sadri,*[32] *darakht-bichoti, karwan, kayali, baqayat, ekotri, charai, dandha,*[33] *nalbant,*[34] *baitul-māl,*[35] *batvih,*[36] etc.

Burden of Agricultural Taxation

Looking at the various categories of taxes, we find that while the *māl-o-jihāt* accounted for the bulk of total rural taxation, the other categories accounted for minor but varying shares. In *pargana* Chatsu, the *sā'ir-jihāt* ranged from 1 to 2%, and in one year to 3%, of the *muwafiq-jama'bandī* ; in Amber it ranged from 3% to 18%, reaching in one year 25%. Perhaps, Amber, being the capital city, had many commercial taxes which were comprehended under the *sā'ir-jihāt*. In other *parganas* the *sā'ir-jihāt* displays fluctuations between the two limits seen in these two *parganas.*

The *siwāi jama'bandī* was not a light imposition by any means. Sometimes it amounted to three-fourths and more of the *muwāfiq jama'bandī*, and even occasionally nearly equalled or exceeded it. But during and after the 1720's the share of the *siwāi jama'bandī* falls sharply in all the *parganas*. This may be seen from Appendix B. This fall is probably to be attributed to the assimilation of a large part of the demand under *siwāi-jama'bandī* to the *māl-o-jihāt*. This is particularly

29. A tax levied on shops in a *bāzāār* either in money or in kind as the rent of the ground on which their shops or stalls are situated. It was also called as *hat bazari.*
30. This tax was levied exclusively in *khāliṣa* villages.
31. Village expenses incurred for payment of various perquisites *shahnagī, payādgī, chaudhurī, farah potakhāna, wāqia-navis* etc. (*Arhsatta* Lalsot). This was collected by the *patel* and the *paṭwārī*.
32. Levied upon *Luhar* (iron-smith).
33. Grazing tax upon animals.
34. A share or rent paid to the rich peasants or *jāgīrdār* by an ordinary cultivator or belonging to menial class (*chiṭṭhī* Malarna 1793/1736).
35. *Baitul māl* literally meant the (Muslim) public treasury to be used for public welfare. Here it specifically means intestate property which was appropriated by the State. (*Arhsatta*-Malarna).
36. Share on the cultivation (*Amber records*, 1744 A.D.).

illustrated by the absolute figures for Lalsot and Chatsu where the *māl-o-jihāt* increases while the *siwāī jama'bandī* falls.

In giving details of the various taxes, we may sometimes seem to have lost the wood for the trees. But the discussion is necessary for understanding the nature of the taxation system, and for distinguishing between the taxes falling upon agriculture and other interests and activities. The amount collected under non-agricultural heads has been found to be far less significant than on the agricultural, as may well have been expected.

With the available information, it is possible to work out, in rough terms, the total burden of taxation on agriculture. We may take the accounts for some villages of *pargana* Naraina in 1730 as an illustrative sample.

The assumption is that the share of each crop taken in every village was the same as in the *pargana* as a whole. This is necessary because whereas for the whole *pargana* the share of the peasants (*ra'iyat*) out of the total produce is specified, it is not specified for the villages separately. Taking the value of the state's share of each crop collected in the village, and applying it to the ratio of the total produce worked out from figures for the whole *pargana*, we can determine the total value of the agricultural produce of each village.

We have then added to the *māl-o-jihāt* the figures for the other taxes, cesses, etc., listed under *sā'ir-jihāt* and *siwāī-jama'bandī*, excluding only such as did not fall on agriculture. To them has been further added *bhom*, the "bhumīas' cess", whose nature has been discussed in Chapter VI. Taking the total amount of tax so determined, we have then worked out the proportion it bore to the total value of produce calculated in the manner explained in the previous paragraph.

The results, showing the extent of all taxation in relation to the total agricultural produce in selected villages, are set out below :

Villages	*Māl-o-jihāt Taxes as % of Produce*	*Total Taxation as % of Produce*
1. Kadawli	36.02%	46.78%
2. Kachnaria	34.39%	44.30%
3. Khawajpura	36.88%	47.28%
4. Goji	33.34%	44.17%
5. Gopinathpura	33.47%	44.13%

On the whole, the total taxation amounts to above 44% of the produce. Thus in addition to the principal land tax accounting for more than a third of the produce, a further tenth of the produce had also to be surrendered directly or indirectly by the peasants. Normally, then, a little less than half of the value of agricultural produce was accounted for by taxation.

But as has been said earlier (Chapter IV), the burden of taxation varied within the agricultural community. Thus the *ra'īyatī*, who formed the bulk of the agricultural population in a village and did not belong to the category of superior right-holders, had to pay land revenue at the higher rate of one-half.

The evidence suggests very strongly that the Mughal revenue system (and the scale of revenue demand sanctioned under it) prevailed in Eastern Rajasthan. Bernier said that the peasants were more lightly assessed in the territories controlled by the *rājās* and in consequence fled to them to escape from the oppression of the Mughal governors.[37] But this would not seem to be applicable to Eastern Rajasthan, where any deviations from the standard Mughal system are hard to discern.

37. Bernier : *Travels of the Mughal Empire* Eng. tr., 2nd, ed. revised by V.A. Smith, p. 205.

APPENDIX A

Taxes Under Sā'ir-Jihāt

1. *Patwāra or Patwaro* : On account of fees of the *paṭwārī taka* 12½ to *taka* 15 per hundred rupees. *Arhsatta* Amber and Bahatri. The statement "*patwāro* (*patwāra*) *patwāri ko Do*" (Let the '*patwāra*' be given to the *patwārī*) occurs in Amber records. (Also see '*arẓdāsht Chait Sudi*, 3, 1645 A.D.).
2. *Pānī* : Tax on irrigation at the rate of three per hundred rupees (of land tax). The water used for irrigation was either by *nāla, talāb*, river or well (*Arhsatta* Lalsot *VS* 1739/1682 and *Chatsu VS*, 1788/1731).
3. *Chailli* : A tax on goat. The rate ranges from *takas* 6 to 9 per head. Sometimes the 'old *chailli*' and 'new *chailli*' are separately given with different rates. (*Arhsatta pargana* Bahatri).
4. *Māpa* : Tax, at 2½% of *mal* (*Māpa Māika*). The following example from *Arhsatta* Amber, 1741 indicates the *māpā* rates applied in the case of different crops cultivated.

Crop	*Kharīf area*	*Rate in rupees*	*Amount*
Varh (sugarcane)	3187.4.0	0.8.0	1594.10.0
Vani (cotton)	4976.11.0	0.4.0	1245.6.2
Tobacco	91.16.0	0.8.0	46.11.0
Kharij (miscellaneous)	—	—	67.8.2

Rabī'

Crop	*Area*	*Rate in rupees*	*Amount*
Tobacco	2.1.0	0.8.0	1.7.0
Kharij (miscellaneous)	—	—	67.8.0

The same *dastūr* was applied in other villages in the *Khāliṣa*.

5. *Māpa-Rāhdārī or rāhdārī* : This tax is listed under *sā'ir-jihāt* as well as *siwāī-jama'bandī* but appears more usually under *siwāī-jama'bandī*. A toll tax on *banjāras*, in both *khāliṣa* and *jāgīr* lands ; this was exacted by the various authorities controlling the routes. The tax was remitted in the Amber territories in 1712 A.D. when prices were unusually high, but it was reimposed in 1713 A.D. (*Chiṭṭhī* dated *Miti Mangsir Vadi* 5, *VS* 1770/1713).
6. *Kayalī* and *Tulāi* : Tax on account of the fees paid to *kayāl* (weigher), when the grain was sold to the merchants. The rate of tax was As. 10 per hundred rupees of *māl* (*Arhsatta pargana* Chatsu and Dausa).

   There is also a tax in *siwāi jama'bandi* designated *peshkash kayālī*. When Dungar, Prahlad and Ratan Singh *mahājan*, *qaṣba* Malarna, called upon the merchants to sell the grain in the market, a duty of *peshkash kayālī* ('weighman's offering') was taken from the merchants (*arhsattas*).
7. *Ghughri Shahna* : On account of the grain presented to the *sehna* (Pers. *shahna*), whose duty was to watch over the grain in a village. He was entitled to take 15 *man* of grains as *shahnagī*. If there were four *shahnas* in a village, 60 *man* of grain were given to them. The rest of the tax realized was desposited as state share under the heading of *ghughri shahna* (*arhsatta pargana* Malarna, 1740 A.D).
8. *Juro* : Tax on a pair of bullocks drawing the plough.
9. *Jhuprī* : Tax on cottages. The rate was As. 10 per *jhuprī* or hut. This is found in all the *parganas*. In *pargana* Naraina the rate varied from As. 8 to As. 12 per hut. On the *jhuprīs* of the *begārīas* (the depressed castes, subject to *begār* or forced labour), the rate was As. 5 only. In *pargana* Naraina (1730 A.D.), the total number of *jhuprīs* on which the tax was realized, was 823, of which 14 huts beloged to the *begārīas*.
10. *Ghiwai* (*byāẓ*) : On account of interest given to the *mahājan* or *banīa*,, *bohra* or *sāhūkār* (moneylender).
11. *Rozīna ṭappādār* : Daily perquisites of the *ṭappādār*. The rate under *ẓabṭi* was *takas* 8 per hundred *bīghas*, and in

jinsī, *takas* 2 on one hundred *mans*. The *ṭappādār* was also paid on a monthly basis (*Arhsatta*,*pargana* Bahatri).

12. *Rozīna* (lit. daily tax.) : According to the *dastūr* the rate was *takas* 29 ; but according to *siwāī-dastūr*, it was *takas* 14.50 in *baṭā'ī*. Perhaps the tax imposed to meet charges of the revenue officers and their party during their stay in a village.

13. *Farhi Quṣūr* : Lit. tax to meet loss. Perhaps this was intended to cover loss of crop/grain collected in land revenue. It also seems to have comprehended fines (*Arhsatta* Lalsot).

14. *Pichotri* : A levy of 5% on the *māl*. When applied in *ẓabṭī* the rate was Rs. 5/- per hundred rupees of tax charged on *aṣlī bīgha*. The exact reason for this impost is not clear (*arhsatta* Bahatri, 1691).

15. *Sarhi* (Pers. *sharah*, rate) : This tax was apparently connected with *ẓabṭi* and was levied at *takas* 1.50 per rupee in *pargana* Bahatri. But in Chatsu, the rate was *takas* 4.75 per rupee. *Sarhi jarībāno* (rate of measurement) was put at *takas* 31 on 100 *bīghas* (*arhsatta* Bahatri and Chatsu).

16. *Ghiv*, *ghiv mahājan*, *ghrit banik* or *bania* : A cess on *ghee*.

17. *Teḥṣīldāri* : Collection. According to *dastūr-ul 'amal* the rate was *takas* 25 per 100 rupees (*arhsatta* Malarna).

18. *Sarh athotri* (Lit. 8½) : A cess in *ẓabṭī* amounting to Rs. 8, as 8, levied on a hundred *bighas*.

19. *Ekotri* (Lit. one) : Re. 1 levied on 100 *bīghas* under *ẓabṭī*. This tax also appears under *siwāī jama'bandī*. It was also imposed on the *patwārī* upon the sale of grain (*tulāi*) at as. 12 and 2 paisa and the merchant at as. 10 and 2 paisa (on Rs. 100/- of sale price). (*Amber Records*, 1678 A.D.).

20. *Nirotri* (Lit. nine) : It was levied at the rate of Rs. 9/- per hundred *bīghas*. If levied on the *tan* (nominal assessment) of a village assigned in *jāgīr*, the rate was Rs. 9/- on Rs. 100/- of the total assessed revenue. Thus also *sarh-chirotri* (Rs. 6½), *sarh-trirotra* (Rs. 3½), *athotra* (Rs. 8/-), *pichotra* (Rs. 5/-) and *ponotra* (as. 12/-).

21. *Go-sumārī* (Pers. *gāu-shumārī*). Cattle tax, rates varying from *pargana* to *pargana*. In Phagi or Phagui (1698) the rates were :

1 rupee on 10 cows.
,, ,, on 14 beasts (*dhor*, prob. bullocks).
,, ,, on 5 milch buffaloes.
,, ,, on 7 male buffaloes.
,, ,, on 27 heads of goat/sheep (*chailli*).

APPENDIX B

Table I

Size of Muwāfiq Jama'bandī (M) and Siwāi Jama'bandi (S)

1. As per cent of Total Demand

	Pargana Amber			*Pargana Bahatri*	
	M	*S*		*M*	*S*
Year	*Jama'bandi*	*Jama'bandi*	*Year*	*Jama'bandi*	*Jama'bandi*
1688	76	24	1665	91	9
			1669	83	17
1689	75	25	1684	80	20
1690	61	39	1686	89	11
1712	24	76	1688	68	32
1715	57	43	1689	60	40
1716	56	44	1691	38	62
1718	54	46	1696	70	30
1720	75	25	1697	62	38
1723	81	19	1706	75	25
1724	63	37	1708	71	29
1725	84	16	1710	70	30
1728	15	85	1716	55	45
1729	37	63	1717	34	66
1733	71	29	1710	64	36
1734	74	26	1720	67	33
1736	86	14	1723	67	33
1737	88	12	1724	70	30
1744	89	11	1725	75	25
	Chatsu			*Dausa*	
1708	63	37	1712	61	39
1709	66	34	1713	45	55
1710	77	23	1715	40	60
1711	76	24	1736	89	11
1712	58	42	1737	90	10
1714	55	45	1738	92	8
1716	63	37	1739	92	8
1717	51	49	1741	89	11

1723	80	20	1742	89	11
1724	81	19	1743	91	9
1726	78	22	1744	94	6
1730	84	16	1745	82	18
1731	76	24	1746	86	14
1733	82	18	1747	93	7
1735	91	9	1750	95	5
1737	92	8			
1738	89	11			
1741	89	11			
1742	88	12			
1744	90	10			
1745	89	11			
1748	94	6			
1749	94	6			

Pargana Lalsot			*Pargana Malarna*		
1687	80	20	1690	77	23
1713	67	33	1711	85	15
1714	53	47	1712	86	14
1715	38	62	1713	78	22
1717	26	74	1714	77	23
1718	46	54	1715	79	21
1719	31	69	1716	65	35
1720	58	42	1719	68	32
1721	61	39	1723	83	17
1722	64	36	1726	75	25
1726	68	32	1727	81	19
1731	70	30	1728	79	21
1733	67	33	1729	80	20
1734	60	40	1730	80	20
1737	88	12	1732	73	27
1739	87	13	1736	74	26
1740	88	12	1737	76	24
1741	85	15	1738	96	4
1745	83	17	1743	96	4
1746	92	8	1744	87	13
1747	87	13	1745	83	17
1748	80	20	1748	90	10

II. In absolute figures (Selected, *parganas* only)

Pargana Amber

Year	*Māl-o Jihāt*[38] *Rs.*	*Takās*	*Siwāi-Jama'bandī* *Rs.*	*Takās*
1720	1,60,337	—	51,208	—
1723	1,46,139	—	1,589	—
1725	4,43,442	1,49,870	1,18,897	4,06,839
1734	1,42,491	51,503	1,10,399	11,035
1737	4,70,554	1,19,020	42,991	29,098
1744	2,89,462	1,87,484	29,521	8,855

Pargana Lalsot

Year	Rs.	Takās	Rs.	Takās
1721	44,595	—	27,636	—
1731	77,399	—	882	425
1733	60,044	—	27,136	22,999
1737	85,099	—	8,145	1,867
1739	71,838	—	9,196	7,568
1740	72,746	—	8,556	4,862
1746	52,657	—	7,396	1,290
1748	56,809	—	100	—

Pargana Chatsu

Year	Rs.	Takās	Rs.	Takās
1723	1,61,798	188	37,978	8,481
1724	1,44,035	—	30,721	8,156
1730	1,84,109	—	30,075	61,158
1735	2,18,020	—	17,533	8,646
1737	2,54,034	—	16,027	7,304
1744	1,07,238	—	12,045	2,433
1749	1,47,431	7,065	8,331	1,458

Pargana Malarna

Year	Rs.	Takās	Rs.	Takās
1723	1,27,252	—	20,158	9,108
1728	1,49,888	—	2,678	6,474
1738	81,055	—	1,766	3,070
1743	88,139	—	1,366	4,792
1748	77,068	—	—	4,838

38. The value of the *takās* per rupee has been given in the *arhsattas*. For the details see Appendix B-2, Also see *The Cambridge Economic History of India*, *Vol. I C. 1200-C. 1750* eds. Tappan Ray Chaudhuri and Irfan Habib, Table 10, p. 310.

CHAPTER VIII

Land Revenue Administration

The Kachhwaha rulers of Amber in the capacity of *jāgīrdārs* were entitled to assess and collect the revenue from lands assigned to them by the Imperial administration. But, at the same time, they were required to conform to the Imperial regulations. These regulations were theoretically binding on the assignees, but their execution depended upon the effectiveness of the central authority ; and when it began to weaken after the death of Bahadur Shah, Jai Singh Sawai and his successors started enforcing their own rules, and regulations in their territories. These were, however, not materially very different from the Mug͟hal pattern.

The Amber rulers on their own assigned *jāgīrs* to their servants and officers. Such *jāgīrs* were administered through the *gumāshtas* (agents) of the *jāgīrdārs*, and the exercise of the ruler's authority in their management was limited. But this was not the case with the administration of the remaining territory, that is, the *khāliṣa* (the portion from which income was collected directly for the treasury of Mahārāja). This was principally managed by the following officials.

Dīwān

The *dīwān*, usually designated *rāj*,[1] was the chief revenue officer. He was the direct representative of the Mahārāja in the

1. In the *Amber records*, which consist of various kinds of *chiṭṭhīs*, the *dīwān* is addressed by the *pargana* officials as *rāj*, literally, the state.

financial and administrative affairs of the *rāj* (state). The appointment of the *dīwān* was directly made by the Mahārāja, and held office during the latter's pleasure.[2] There were two kinds of *dīwān*, the *dīwān-i-tan* (in-charge of *jāgīr* assignments) and *dīwān-i-desh* (*waṯan*).[3] This is the reason why we often find names of two persons in the documents working as *dīwān* (see Appendix to this Chapter). But sometimes only one person was vested with the powers of the two offices.[4]

The *dīwān* was required to manage the assessment and collection of land revenue, from the *parganas*, each of which had its own officials.[5] He had to keep himself well acquainted with the affairs of the *rāj*, particularly with the activities of the different, *pargana* officials such as the *'āmil*, *amīn*, *chaudhurī*, *qānūngo*, *potdār* (Pers. *fotadār*), etc. They were his media of contact with village officials such as the *paṭel*, *muqaddam*, *paṭwārī*, and other petty officials. The *dīwān* was required to assure himself that none of these officials exacted illegal or forbidden cesses or such as had been remitted by the *rāj*.[6] He used to check the records maintained by the *paṭwārīs* in order to prevent suppression of facts by the *pargana* officials.[7] If any *'āmil* or other *pargana* official was found guilty of misappropriation, the *dīwān* was required to take necessary action. He was required to keep an eye on the activities of the *mahājans*,[8]

2. *Amber Records*, *VS* 1724/1667.
3. *Amber Record Vadi* 7 *VS* 1744/1687. The term *dīwān-i-tan* was a direct importation from Mughal administration. *Dīwān-i-desh* would seem to correspond with the Mughal *dīwān-i-Khāliṣa*.
4. Ibid. We also come across a *dīwān* with the designation *dīwān-i-huzūrī*.
5. *Chiṭṭhīs* of *pargana* officials sent to the *dīwān*.
6. *Chiṭṭhī VS* 1803/1746. Instructions were sent to the *'āmil* and the *amīn* to instruct the *patel* not to demand *malba* from peasants. *Chiṭṭhī* of *dīwān* to *'āmil* and *amīn*, *qaṣba* Paparda, dated *Miti Jeth Sudi* 15, *VS* 1802/1745 about the exemption of *begār* and *jhupri*.
7. *Muwāzana Kalan* and *khurd*, *khasra* and other kinds of village papers were sent to the office of *dīwān*, duly signed by the village officials.
8. *Chiṭṭhī Miti Sawan Sudi* 2, *VS* 1762/1705. *Dīwān* Kishore Das wrote to Mahārāja Jai Singh Sāwāi that he had called the *mahājan* and instructed him not to raise the prices of the *jinsī* (foodgrains) in the days of drought, that the price should remain uniform throughout the city, and that he should not store grain.

since the *mahājans* played an important role in the sale of food-grains, and in the collection of revenue. He entertained complaints lodged by the village officials such as the *paṭel* and the *paṭwārī*, and also from other officials as well from the *ra'iyatī*.[9]

The *dīwān's* jurisdiction extended to all the various categories in which land was distributed, viz., *khāliṣa*, *jāgīr*, *punya-udik* and *in'ām*,[10] *peshkash bhomī* and *ijārā*. The rights regarding the renewal of *paṭṭas* of different types were vested in him. All officers serving in a *pargana* were required to send copies of records and accounts to the office of the *dīwān-i-huzūrī*, duly signed or attested by the concerned official.

The appointments of *pargana* officials such as the *'āmil* and the *amīn*, *potdār*, *chaudhurī*, *qānūngo*, etc., were made by the *dīwān*.[11] The revenue collectors (*'āmils*) and assessors (*amīns*) worked under the direct supervision and control of the *dīwān*. He was required to see that all papers such as the rolls of *jama*,' *jama'-o-kharch*, *awārija*, *yaddāsht* etc., were sent to him regularly. He was exhorted to increase cultivation and make provisions for seeds and oxen (*bij* and *balad bail*) to improve cultivation in the *pargana*. He was simultaneously urged to prevent the oppression of *ra'iyatī* by officials, *jāgīrdārs* and the *zamīndārs*.[12] He issued instructions to the *pargana* officials for special measures regarding cultivation during famine days.[13] He was required to transmit all orders of the Mahārāja to the

9. Documents of this kind are too numerous to be cited.
10. See a large number of *sanads* of *punyaudik* and *in'ām*. *Chiṭṭhī VS* 1773/1716 ; *Chiṭṭhī Sudi* 12 *VS* 1768/1711 ; *Sanad qarār Miti Sawan Vadi* 10, *VS* 1805/1748 ; *Chiṭṭhi Miti Sudi VS* 1804/1747.
11. *Chiṭṭhī Asarh Sudi* 12 *VS* 1761/1704 and *chiṭṭhī of dīwān Miti Chait Vadi* 6, *VS* 1801/1744.
12. *Parwāna Miti Asarh Sudi* 8, *VS* 1783/1726 ; *Chiṭṭhī* of *dīwān* addressed to *faujdār* and *'āmil pargana* khohri *VS* 1804/1747 ; *Chiṭṭhī VS* 1799/ 1742 : The Askaran Brahman of *mauza* Chharhera, *pargana* Phagi or Phagui, wrote to the *dīwān* that he had cultivation of his own (*gharoohala*), and *vat Chautho* (one-fourth) had been fixed upon him as the state demand for which he possessed a *sanad* of the *dīwān desh*. But the *'āmil* was making trouble in regard to the *vat* (share). The *'āmil* was instructed to realize revenue according to the *sanad*.
13. *Chiṭṭhī Miti Jeth Vadi* 8, *VS* 1722/1665 *pargana* Naraina.

pargana officials. He could also transfer *pargana* officials on the basis of complaints received against them either from the *ra'iyatī* or the *zamīndārs*.

The *dīwān* was in a sense the custodian of the ruler's treasury. He kept a close eye on the *potakhāna* (treasury), and checked the accounts, particularly, of *jama,' bāqī* (balance) and *baqāyā* (arrears). He was obliged to ensure that no money was spent or appropriated without proper sanction for its disbursement. There was a special official, the *mustaufī*, whose business it was to audit the accounts of the *'āmil* and the *amīn*.[14]

Naturally, the *dīwān* had to keep a fairly large secretariat. There were separate offices to receive Persian and Hindi correspondence. In each office there were several *muharrirs* (clerks).

Pargana Functionaries : 'Āmil and Amīn

A number of villages formed a *pargana*, which was a fiscal as well as a territorial unit. *Pargana* in our documents is also sometimes referred to as *maḥal* or *ẓila*.[15] A *pargana* could consist of *ṭappas*[16] (a group of villages). The *ṭappa* was usually named after the principal village situated within it.

The most important land-revenue officers in a *pargana* were the *'āmil* and the *amīn*. These two officers had to work under the direct supervision of the *dīwān*. Our documents show that the offices of the *'āmil* and the *amīn* were separate from each other and two different persons were usually appointed to these posts.[17] But sometimes, the two offices could be combined in the same person. For example, Nanig Ram was appointed in 1736 as *'āmil* and *amīn* in *pargana* Sawai Jaipur.[18]

14. *Chiṭṭhī Miti Bhadva Sudi* 2, from *dīwān* Rai Narain Das Kirpa Ram to *Sah* Ramanand, *'āmil pargana* Khatoo and Amarsar, instructing him to look after the interest of the state and the welfare of the *ra'īyati* ; he was further to send the papers to *dīwān-i-hūzūri* and *mustaufī*.
15. *Chiṭṭhī Miti Pos Sudi* 1, *VS* 1774/1717 ; *Chitthī Miti Jeth Sudi* 12, *VS* 1775/1718 ; *Chiṭṭhī Miti Jeth Sudi* 15, *VS* 1788/1731 ; *Chiṭṭhī VS* 1767/1710.
16. *Ṭappa* was a fiscal unit of a *pargana*.
17. See Appendix for the names of *'āmils* and *amīns* of different *parganas*.
18. *Arhsatta pargana* Sawai Jaipur, 1736.

The word *'āmil* seldom occurs in the official documents of Eastern Rajasthan from the middle of the 17th century up to the 1680's.[19] Instead, the word *kārkun* is used. It may be presumed that the duties of the *kārkun* and *'āmil* were identical or that the *amīn* was entrusted with the work of the *'amīl* during this period. After 1690, roughly from the time of Mahārāja Bishen Singh, the office of *kārkun* ceases to be mentioned in the documents. Most probably, it was replaced by the office of the *'āmil.*

The *'āmil* was entrusted with the task of ensuring cultivation of all arable land and the collection of the assessed revenue from the cultivated lands.[20] He had to take the help of the *tappadār*, *chaudhurī*, *qānūngo*, *shahna* and other village officials.[21] All the letters of receipt and disbursement such as *arhsattas* had to be duly signed by the *'āmil* and the *amīn.*[22] The papers of *muqarrara jama'* (fixed revenue demand), *ḥal ḥaṣil* (revenue realized in the current year), *minzalik* (expenditure), and *bāqi* (balance) were attested by these officials individually. The *jama'* papers were also duly signed by the *chaudhurī* and the *qānūngo.*[23] To ensure cultivation of all arable land, the *'āmil* recommended to the government the advance of *taqāvī* loans, or himself managed to obtain the same from the local *bohras* or money-lenders. He was also enjoined to expand the area under cultivation by getting more land, especially the *ujar* (waste land), brought under the plough. In this work, he received instructions from the *dīwān* or sometimes direct from the Mahārāja ; and he was vested with special powers. He could allow the peasant the option of having the revenue levied by *ẓabṭī* or

19. See Appendix, of this Chapter.
20. *Chiṭṭhī Miti Asoj Sudi* 7 *VS* 1746/1689.
21. *Chiṭṭhī US* 1782/1725 of Kanhaji and Bhim Raj (*amīn* and *'āmil* respectively) to *dīwān* Kishor Das and Tara Chand ; *Chiṭṭhī VS* 1780/1723 of Bakht Ram and Khushali Ram (*amīn* and *'amil*) to *dīwān* Narain Das Kirpa Ram.
22. See *arhsattas* of different years for different *parganas*.
23. *Arhsatta pargana* Malarna *VS* 1722/1765. Apart from the seal of *sāh* Rūp Chand (*amīn, pargana* Malarna) the signatures of *chaudhurī* Hayat Khan and Wali Khan and *qānūngo*. Pohkar Rūp Rai Ratan Singh are impressed on the *jama' mujmil* and *bāqi, etc.*

kūt, (*koont*) or paying in cash or kind.[24] He was authorised to issue new *paṭṭās* of *vat-vih* (share in cultivation) at concessional rates, ranging from one-third to one-fourth to these cultivators who brought *parat* (fallow) lands under the plough.[25]

The *'āmil* was required to see that he obtained correct reports from the village officials about the cultivable land and the quality of crops.[26] He regularly forwarded to the *rāj* information regarding famines and other natural calamities which affected the cultivation ; and he implemented the measures prescribed by the *dīwān*, e.g., provisions for *hal* (plough), *bij* (seed) and *bail* (oxen) for the *paltī* (cultivator). Instructions were also given by the *'āmil* to the *chaudhurīs qānūngos, paṭels* and *patwārīs* to search and rehabilitate the *ra'iyatī*, who had abandoned cultivation due to famine and to provide them all facilities for cultivation.[27] He was required to check the work of the *chaudhurī* and the *qānūngo* and keep strict vigilance over them as well as other village officers. He was expected to prevent undue exaction or oppression and to see that his instructions were properly implemented.[28] He had to keep watch on the grain merchants (*banjāras*) and protect their interest against encroachments by the *zamīndārs* and *bhomīas*.[29] It was again, his duty to see that the grain was sold to merchants from the *kothās* (store-houses) before the commencement of the rainy

24. *Chiṭṭhi VS* 1782/1725.
25. *Chiṭṭhi Miti Jeth Vadi* 8, *VS* 1722/1665 ; *Chiṭṭhī VS* 1707/1650 of Sewak Shyam Sahi and Ghulam Sundar to Mahārāja. This *chiṭṭhī* was a reply to a *parwāna* sent by the Mahārāja to these officials.
26. *Chiṭṭhī VS* 1724/1667. Ajit Das and Lakshmi (*'āmil* and *amīn*) informed *diwān* Kalyan Das that they are sending him the papers of *takhmīna bīghawar* and *ek harfi* for *jinsī*. Also see *Chiṭṭhī Miti Sawan Vadi* 10, *VS* 1723/1666.
27. *Chiṭṭhī VS* 1722/1665 from Sri Ram to *dīwān* Kalyan Dass. *Chiṭṭhi Baisakh Sudi* 3, *VS* 1722/1665 from Kesho Das to *dīwān* Kalyan Das *Chiṭṭhī VS* 1753/1696 from Girdhar Das Manrūp (*Kārkun* and *'āmil* respectively) to Mahārāja Bishan Singhji.
28. *Chiṭṭhī Miti Pos Sudi* 7, *VS*, 1721/1664. *Chiṭṭhī VS* 1787/1730 about the arrest of a *paṭel* by the *'āmil* who was imprisoned for four days. *Chiṭṭhī Jeth Sudi* 5, *VS* 1724/1667. *Chiṭṭhī Miti Baisakh Sudi* 13, *VS* 1722/1665.
29. *Chiṭṭhī Pos Sudi* 11, *VS* 1754/1696 of Harnam (*'āmil*) to Mahārāja Bishan Singh. *Chiṭṭhī Miti Jeth Sudi* 13 *VS* 1775/1718.

season.[30] He had often to fix the amount of *peshkash* levied upon the *zamīndārs*. The settlement of disputes among the *jāgīrdārs*, *zamīndārs*, *ra'iyatis* and *ijārādārs* with regard to the fixation of their shares came within his jurisdiction.[31] The *'āmil* was to protect the *ra'iyatī* and the *jāgīrdār* from the oppression of "*zor-ṭalab*" *zāmīndārs*.[32] Usually, the *'āmil* and the *amīn* stood surety (*māl-i zāmini*) on behalf of an *ijārādār*.[33]

30. *Chiṭṭhī Miti Asarh Vadi* 6, *VS* 1722/1665.
31. *Chiṭṭhī* of *dīwān* dated *Jeth Vadi* 2, *VS* 1804/1747 to the *'āmil* of *pargana* Sawai Jaipur : "A complaint was lodged by the *gumāshta* of Sukh Singh *Khichi*, the *jāgīrdār* of village Chimanpur, situated in Charura, *ṭappa* Ramgarh, *pargana* Sawai Jaipur that the *ra'īyat* of this place do not recognise the *jāgīrdāri* rights."
Chiṭṭhī VS 1749/1692. *Chiṭṭhī Miti Sawan Sudi* 10, *VS* 1802/1745.
Chitthi VS 1801/1744 of *dīwān* to the *'āmil* : "Jai Singh Kumbhani, *jāgīrdār*, village Udaipur *ṭappa* Patan, Sawai Jaipur had lodged a complaint "that the *'āmil* has given the above-mentioned village in *ijārā*". The *'āmil* was instructed to see to it that the dispute between the *jāgirdār* and *ijārādār* should be settled properly, and to fix the *haṣil* of *syalu* (*Kharīf*) after proper scrutiny of *khasra* papers.
Chiṭṭhī Miti Bhadva Vadi 7, *VS* 1807/1750 : The *'āmil* was instructed to settle a dispute between the *jāgīrdār* and *pālatyā* (cultivators).
Chiṭṭhī Asarh vadi 8, *VS* 1781/1725.
Chiṭṭhī Miti Chait Sudi 9, *VS* 1722/1665. The *'āmil* informed *dīwān* Kalyan Das that he had fixed one-fourth (*vat chautho*) to the *ra'iyatī* in *banjar* land, whether *khudkāsht* or *pāi-kāsht*.
Chiṭṭhī Asarh Vadi 8, *VS* 1782/1725. The *patels* of *mauza'* Peelwara and village 'Sumel of *pargana* Udai, complained to *dīwān* Narain Das Kirpa Ram regarding a land dispute. The *dīwān* referred the matter to Jethmal and Akhari Ram (*'ā'mil* and *amīn* respectively). Then the *faujdār*, *'āmil* of *parganas* Malarna and Udai, *paṭels* of five villages of Malarna and Udai, were called and accepted as *panch* ; and a decision was taken before more than one thousand people of the villages.
32. *Chiṭṭhī Miti Asadh Sudi* 13, *VS* 1740/1683. The *faujdār* is helping the *'āmil* to suppress the *zamīndārān- i zorṭalab*. (*Chitthī VS* 1721/1664). Maha Singh wrote to *dīwān* Kalyan Das that "the *zāmīndār* of village Giglana was not giving *'amal* (jāgīrdāri rights) to Khem Karan, the *jāgīrdār* of the aforesaid village. The *jāgīrdār* went to *bakhshī* Ratan Singh to use force to eject the *zamīndār*. Consequently, the *bakhshī* along with a force of 3000 *asawār* (horsemen) attacked the aforesaid village 'on *Pos Sudi* 12, Monday. Many people were killed. Ultimately, he succeeded and the *jāgīrdār* was given the charge of that village.' (*Chiṭṭhī* dated *Miti Bhadva Sudi* 15, *VS* 1749/1692.).
33. *Arhsatta pargana* Malpura *VS* 1782/1725. Sanghi Laxmi Dass and

The *'āmil* and the *amīn* were both responsible for the collection of *baqāyā* (arrears) in their *pargana* ; the *dīwān desh* demanded *muchalka* (paper of agreement) from the *'āmil* and the *amīn* of each *pargana* owing responsibility for realization of the arrears.[34] It was the duty of the *'āmil* to see that the papers of *baqāyā* were duly signed by the *chaudhurī* and the *qānūngo* of the *pargana*[35], and the amount of *ijāra* was verified and duly signed by the *qānūngo* and the *paṭel*. He was authorized to grant *takhfīf* (reduction) in the *baqāyā*.[36] He was required to get the *dahsāla dastūr-ul 'amal* (revenue and tax rates for ten years) and send it to the *dīwān's* office.[37] Instructions were given to the *'āmil* to adopt such methods that the *jama'* of the *pargana* might show an increase every year.[38] The villages which he thought had a *gunjāīsh*) margin of profit in revenue in the areas assigned) were to be set aside by him to be retained in *khāliṣa* instead of their being farmed out for a lump sum or given in assignment.[39] He used to appoint the *shahna*,[40] and was required to see that the rights of *shahna* and *paṭeli* are duly protected in a village.[41]

Gulab Chand (*amīn* and *'āmil* respectively) of *parganaa* Toda Rai Singh stood surety for Churamani (*ijārādār*) of village 'Kukri, *pargana* Toda Rai Singh, for Rs. 978.11.0 Usually, they '('*āmil* and *amīn*) were *mahajans*.

34. *Parwāna* from Mahārāja to Vidhyadhar, *dīwān desh VS* 1782/1725. (*Chiṭṭhī Miti Kartik* 9, *VS* 1739/1682.) The *'āmil* reported to the *dīwān* "that the *paṭel* and *patwārī* are thieves since they keep the *baqāyā* of the *sarkār*." *Chiṭṭhī Miti Mah Sudi* 3, *VS* 1739/1682.
35. *Chiṭṭhī Miti Asoj Sudi VS* 1724/1667 ; *Chiṭṭhī VS* 1788/1731 ; *Chiṭṭhī VS* 1798/1741.
36. *Baqaya Qaṣba* Sanganer, *Miti Kati* (*karitik*) *Vadi* 1, *VS* 1758/1701.
37. *Chiṭṭhī Miti Sawan Vadi* 11, *VS* 1801/1744. Radha Kishan informed *dīwān* Murlidhar "that as desired by you he is sending the *dehsāla dastūr-ul 'amal* to the office of the *dīwān*.
38. *Chiṭṭhī dīwān Miti Bhadva Vadi* 13, *VS* 1761/1704 to the *'āmil* of Sanganer. *Chiṭṭhī Miti Asoj Vadi* 4, *VS* 1807/1750. The *'āmil* instructed the *paṭel* that the *ra'īyati* should not be allowed to keep the land as *parati* (fallows). (*Chiṭṭhi Miti Jeth vadi*, 8 *VS* 1722/1665, *pargana* Naraina).
39. *Chiṭṭhī*, *Miti Baisakh Sudi* 8, *VS* 1761/1704.
40. *Chiṭṭhī VS* 1805/1748 ; (*sanad*) from *dīwān* Rai Narain Das to *'āmil* Sah Har Narain, *pargana* Sawai Jaipur.
41. *Chiṭṭhī Miti Bhadva Vadi* 10, *VS* 1806/1749, ; *Chiṭṭhī Miti Pos Sudi* 7, *VS* 1801/1743 ; *Chiṭṭhi Miti Chait Sudi* 9, *VS* 1722/1665.

The *'āmil* looked after the *in'ām* and *punyaudik* lands, and was authorised to demand the *sanads* on behalf of the *rāj*.[42] It was on the recommendation of the *'āmil* that the *dīwān* used to issue *sanads* to *madad-i ma'āsh* holders. He also granted *paṭṭas* to *varhdārs* (watchmen) who protected the standing crops from animals.

The *'āmil* was supposed to ensure extension of cultivation and enlarge revenue. It was therefore incumbent on him to be concerned with the establishment of new villages, including the act of naming them.[43] He invited *paltīs*, *pā'is* and others to the new village and gave them the necessary assistance in the form of allotments of land, ploughs, seed, etc. *Khud-kāsht* cultivators, who also played a role in bringing the land of newly founded villages under cultivation, were granted *paṭṭas* at concessional rates by him. If there were any territorial disputes among the *jāgīrdārs*, the *'āmil* demarcated the land (*chakbandī*).[44]

The function of the *amīn* was to see that the dues of the *rāj* (*māl*, *sā'ir* etc.) were properly levied on the *ra'iyatī*. His main duty was to ascertain the area of land under cultivation and to assess it according to the existing rules and regulations. He was required to send the assessment and other papers maintained in his office to the revenue department. He was required to inform

An *'āmil* punished a *gunwār* (cultivator) when he misbehaved with a *shahna*. The *gunwār* was imprisoned for a month. Since that was the time of sowing it was decided that whatever he produced in a month should, in lieu of imprisonment, be taken from him as punishment.

42. Copy of a *sanad Miti Sawan Vadi* 10, *VS* 1805/1748 from *dīwān* Narain Dhar Har Govind to *Sah* Har Narain. *Sanad qarar Miti Phalgun Sudi* 7. *VS* 1805/1748 about the grant of a *paṭṭa* of land in *in'ām* to Vidhya Nath in lieu of guarding land from locusts.
43. *Chiṭṭhī VS* 1801/1744 of *dīwān* to purohit Jai Chand and *Sah* Har Narain. *Arhsatta* Malarna 1730.
44. *Chiṭṭhī Sawan Vadi* 3, *VS* 1806/1749. *Chiṭṭhī Miti Asoj Vadi* 8, *VS* 1749/1692. *Chiṭṭhī Miti Baisakh Vadi* 3, *VS* 1804/1749. Amar Singh s/o Roop Singh Kumbhawat had a *jāgīr* worth Rs. 350/- in village Batwara, *tappa* Kakar, *pargana* Sawai Jaipur. The *jama'* of the village was Rs. 3, 350/-. The rest, worth Rs. 2000/-, was assigned to Gaj Singh Kumbhawat. There was some disputed land between Amar Singh Kumbhawat and Gaj Singh Kumbhawat. Therefore, it was decided to measure the land from one side, according to *chakbandī*.

the *rāj* about conditions of rain in his *pargana*.[45] The *amīns* and the *'āmils* were responsible for the circulation of new coins in their respective *parganas*.[46]

Though the *amīn* and the *'āmil* held separate offices, their functions largely overlapped. Whenever there was any complaint about the assessment, the *dīwān* issued instructions in a single letter addressed to both of them ; only very rarely were the two officers addressed separately. Within a few years of the death of Jai Singh Sawat, reference to the *amīn* in the official records begin to fade away.[47] It might well be that the *'āmil* was now performing the duties of *amīn*, or both the offices were commonly held by the same person. This particularly holds true of those territories which were under the direct control of the Mahārāja.

The appointment letters of the *'āmil* and the *amīn* usually classify their remuneration into *alufo* (Pers. *'alufa*) and *lawāzimo* (*lawāzima*).[48] Both these officials received their salary from the *poṭdār* (*foṭadār*) of their respective *parganas*.[49] The salary of

45. *Chiṭṭhī Miti Asadh Sudi* 12, *VS* 1722/1665. *Chitthī VS* 1744/1687. *Chiṭṭhī Miti Sudi VS* 1721/1664. The *amīn* writes to the *dīwān* about the rains in *pargana* Udai : "the *dhan* (paddy) is not good due to scarcity of rain. As soon as the rains fall, the tanks and ponds (*sar-talab*) will be full of water and the cultivation of *dhan* will be possible".
46. *Chiṭṭhī Miti Phalgun Sudi VS* 1721/1664. The *amīns* were instructed by the *dīwān* that the new Alamgīrī coin (*paisa*) should be in circulation in their respective *parganas* and nobody should be allowed to keep the old *paisa* and should be punished in case of being found guilty of retaining the old one. This piece of information is of interest for Mughal monetary history as well, since Aurangzeb minted a new lighter *dām* of copper.
47. *Chiṭṭhī* of *dīwān* Madhuban Das to Sanghi Shri Kalyan Das, *amīn pargana* Malpura. See also *Khulāsatus Siyāq* for reference to separation of offices of *amīn* & *'āmil* under Shah Jahān Br. Mus. MS, Or. 2026, ff 339-349 ; *Irfan Habib, Agrarian System* op. cit. pp. 275-76, 295.
48. *Sanad Miti Chait Vadi* 6, *VS* 1801/1744. The payments of the two officials appearing in accounts were divided under two heads (i) *Dol* or *zāt* (personal), and (ii) *lawāzima* (perquisites). These were generally about equal. The salary of the *'āmil* in *pargana* Malarna, year 1720 was Rs. 1800/- per annum, of which *dol* was Rs. 900, and *lawāzima* Rs. 900. See *arhsattas* of different *parganas.*
49. *Arhsattas* of different *parganas*.

the *'āmil* and the *amīn* used to be normally equal. But the rate was different in different *parganas* depending, of course, on the size and importance of the *pargana*. Usually, the salary was calculated on a yearly basis,[50] and was paid to the *'āmils* and *amīns* according to their period of tenure which could be for four months or six months or for a season. The information on the salaries of the *'āmils* and the *amīns*, as well as the *potdār*, as extracted from various documents, is tabulated below :

Annual Salary of *'Āmil/Kārkun*, *Amīn* and *Potdār* (in rupees)

Year	*Pargana*	*'Āmil/Kārkun*	*Amīn*	*Potdār per month*
1664	Niwai	—	200	18
1732	,,	1,000	—	—
1739-42	,,	912	—	—
1664	Chatsu	1,500 (*Kārkun*)	1,500	—
1694	,,	988	988	—
1725	,,	912	—	—
1730	,,	1,669	1,669	—
1714, 1716	Malpura	800	900	18
1717-19	,,	900	900	18
1720-1722	,,	1,800	1,800	—
1723	,,	1,600	1,600	12
1726-1730	,,	1,600	1,600	12
1732	,,	1,600	1,600	15
1734-35	,,	1,600	1,600	12
1741-43	,,	1,600	1,600	12
1750	,,	1,800	1,800	20
1665	Gaonri	600 (*Kārkun*)	700	—
1716	Mauzabad	1,200	—	18
1730	Malarna	1,400	1,400	—
1731	,,	1.600	1,600	—
1730	Naraina	1,300	1,300	—
1740	Bahatri	800	800	—
1744	Khandela	1,400	1,400	—
1746	Lalsot	1,400	1,400	12
1748	,.	912	—	—

50. *Arhsatta pargana* Chatsu, *VS* 1787/1730 ; *Arhsatta pargana* Malarna, *VS* 1787/1730 ; *Arhsatta pargana* Bahatri, *VS* 1797/1740 ; *Arhsatta pargana* Naraina, *VS* 1781/1730 ; *Sanad pargana* Khandela, *VS* 1801/17744 ; *Chiṭṭhī Mah Vadi* 14, *pargana* Khandela, *VS* 1801/1744.

The appointment of *'āmil* and *amīn* was made by the *dīwān* presumably with the approval, of the Mahārāja. One of the *sanads* of appointment reads as follows : "It is hereby ordered that Shri *Sah* Surat Ram has been appointed as *'āmil* of *pargana* Malarna *Vice* Shri Kirpa Ram. The *'āmil* is advised to look after the interests of the *rāj* and the prosperity and welfare of the *ra'iyatī*. The emoluments and the perquisites (*'alūfa* and *lawāzima*) of the *'āmil* shall be the same as those of his predecessors, which he can collect from the *potdār* of the aforesaid *pargana* from the time of the commencement of his service."[51] A similar document with nearly identical contents was used for the appointment of the *amīn*.[52]

The Appendix contains the names of *'āmils* and *amīns* of different *parganas* appearing in our records. Most of them were *Sahs* (*banīas*) or *mahājans*[53], though Brahmans (*purohit*), Bhayyas (Kayaths or Kayasths) and Khatris also occasionally held these offices. Being *mahājans*, the *amīns* and *'āmils* occupied a key position in the fiscal administration. They probably also gained by standing surety (*māl-i ẓāmin*) for the *ijāradārs*, along with the village *paṭel*.

Chaudhurī and Qānūngo

Two other important *pargana* officials concerned with the land revenue administration were the *chaudhurī aud qānūngo*, whose functions and duties in the seventeenth and eighteenth centuries were practically the same as elsewhere in the Mughal Empire.

The offices of the *chaudhurī* and *qānūngo*, though hereditary, were subject to renewal by the *rāj* (state).[54] Sometimes, however, on the payment of a large sum to the state, the succession could

51. *Sanad qarār Miti Chait Vadi* 6, *Vs* 1801/1744 ; *Chīṭṭhī diwāni qarār Miti Bhadva Vadi* 13, *VS* 1761/1704 ; *Chiṭṭhī Asadh Sudi* 12, *VS* 1761/1704 ; *Chiṭṭhī VS* 1799/1742 : *Chiṭṭhī Miti Mah Vadi* 14, *VS* 1801/1744.
52. *Chiṭṭhī Miti Pos Sudi VS* 1806/1749, regarding the appointment of Khusal Chand as *amīn* of *pargana* Bhadlao.
53. See Appendix ; Also *arhsatta pargana* Malarna, *VS* 1787/1730.
54. *Arhsattas pargana* Chatsu *VS* 1751/1694, *VS* 1787/1730.

be altered by the removal of one incumbent and the appointment of another.[55] An undated *iltamash* (petition) of Gokal Ram Brahman to Māharāja Bishan Singh says, "the petitioner possessed *chaudharīes* of *pargana* Chatsu since the time of Māharāja Bharamal. Now his livelihood (the right of *chaudhuraī*) has been assigned to Jag Ram, a *bania*. But he (Brahman) is prepared to pay one and a half time more than him in case this right is restored to him".[56]

The *chaudhurī* was supposed to render all help to the *amīn* and the *'āmil* in the collection of revenue.[57] The *chaudhurī* was also used as a countercheck to the *qānūngo*.[58] As regards the duties of the *qānūngo*, he was required to collect detailed information about the land revenue and enter the same in his records under different heads.[59] Both the *chaudhurī* and the *qānūngo* were required to prepare the *taqsīm* of different villages.[60] The *chaudhurī* and the *qānūngo* used to prepare consolidated *arhsattas* of different *parganas* on the basis of various papers such as *yāddāsht*, *sanad*, *parwāna* *roznāmcha*, *nirkh bāzār*,

55. *Iltamāsh* (undated), register No. 124, RSA Bikaner. *Chitthī Miti Sudi* 3, *VS* 1799/1742. *Chaudhurī* Mūl Chand and Puran Mal complaining that the rights of *chaudhurāī* had been taken away from them by the *mutaṣaddīs*, and requesting further that the rights should be restored to them according to *dastūr*.
56. *Ibid.*
57. *Chiṭṭhī* of *dīwān* to *chaudhurī* Maya Ram dated *VS* 1799/1742 states that the *gumāshtas* of *jāgīrdār* and *faujdār* Haji Latif Khan of *pargana* Bhadlao complained that due to the scarcity of rain the *'āmil* had collected only a nominal amount of revenue from the *ra'īyatī*; so the *jāgīrdār* was unhappy. It directed him to go and to manage the collections of the *jāgīrdār* according to *dastūr* and to take the help of *paṭel* in this matter, if necessary.
58. *Chiṭṭhī Miti Pos Sudi* 6, *VS* 1721/1661. The *ra'īyatī* lodged a complaint against Bijai Ram, the *qānūngo* for his oppressions. The *chaudhurīs* and *muqaddams* acting as *panch* gave their decision against the *qānūngo* for illegal exactions from the village.
59. See *Chiṭṭhī VS* 1794/1737 of Khushali Ram (*'āmil*) to *dīwān* Vidhyadhar informing him that since the *chaudhurī* and *qānūngo* had gone to *pargana* Pragpur Antela, it was not possible to prepare *jamābandī* in their absence.
60. *Chiṭṭhī Asarh vadi* 8, *VS* 1782/(1725. *Chiṭṭhī Miti Phalgun vadi* 10, *VS* 1721/1664. Gopinath *qānūngo* was instructed to prepare a *taqsīm* of ten years. Such *taqsims* survive in large numbers.

khasra, *parwāna*, and *punya-udīk*, and to affix their signatures on the first page of the document, i.e. over the *muqarrara jama'*.[61]

The *rāzīnāma* (agreement) regarding *ijāra* was also sometimes confirmed by the *chaudhurī* and the *qānūngo*. The *chaudhurī* also stood surety (*māl-i zāmīni*) for the revenue contractors. The occurrence of famine or failure of rain had also to be verified under the signatures of *chaudhurī*, *qānūngo* and *qāzī*.

The *qānūngo* and the *chaudhurī* were entitled to an allowance called *nānkār* out of the revenues. In 1664, it came to 0.82% of the total revenue in *pargana* Gaonri.[62] They were also entitled to *dastūr chaudhurī-qānūngo* from *ra'iyatī* as well as from *ta'aluqadārs'* villages. They used to receive a share in the collection of *muqta'i* (fixed amount of revenue) which was sometimes a half of it, the balance of the allowance due being left with the state. Besides this, they were granted *paṭṭas* of cultivation at concessional rates which usually amouted to one-fourth.[63]

Potdār

There was a separate *potdār* (Pers. *fo ṭadār*) or treasurer for each *pargana* whose functions included receipt of the revenue, safe custody of cash, maintenance of accounts and proper disburse ment of cash.[64] The *'āmil* and the *amīn* were required to draw their salary from the office of the *potdār*, named in their appointment letter.[65] The post of the *potdār* was generally held by a *mahājan*.[66] He had to maintain the *roznāmcha* of daily income and expenditure submitted by *taḥvīldārs* (the collectors) of different departments. He was required to accept all kinds of coins such as *muhr*, rupee, *takā* (or *murādī*) (gold, silver and copper coins respectively). The money could be deposited in the treasury

61. *Chiṭṭhī Miti Asoj Sudi* 13, *VS* 1724/1667 ; the actual *arhsattas* of different *pargana* may also be seen.
62. *Takhmīna pargana* Gaonri.
63. *Dastūr-ul 'amal* and *'amal dastūr* of *pargana* Phagi, 1748/1691.
64. *Arhsatta potdār pargana* Amber *VS* 1720/1663. This gives the names of the *potdārs* in different *parganas* during *VS* 1720/1663.
65. *Arhsatta pargana* Malarna, Naraina *VS* 1787/1730. See *sanads* in these *arhsattas*.
66. See *arhsattas* of *parganas* Niwai, Malpura and Lalsot for different years.

both in Shāhjāhāni and Aurangshāhī rupee. The *potdār* was not authorised to disburse money without the prior sanction of the *dīwān*. It appears that the central treasury was at Amber, and the *potdārs* of different *parganas* were required to send the details of *bāqi* (balance) to the *dīwān*.[67] There are numerous *parwānas* issued by the *dīwān* asking the *pargana potdār* to deposit *bāqi* in the state treasury. The *potdār* was required to send a *ṭalab* (demand) of *bāqi* on *mahājans*, whenever state's money had been deposited with the latter. It was the duty of a *potdār* to ensure that *murādī* or *takā* was exchanged into rupees by the *mahājans* at proper rates.

The *potdār* was supposed to be a *mahīnadār*, i.e. his salary was paid to him on a monthly basis. The salary of the *potdār* was different in different *parganas* under different years, as in the case of the *'āmil* and the *amīn*. His pay has been already tabulated along with that of the *'āmil* and the *amīn* by multiplying his monthly salary by 12 to obtain the annual rate. The rate was naturally much lower than that of the pay of the *'āmil* and the *amīn*.

Ṭappadār

The official appointed over a *ṭappa* by the *dīwān* on the recommendation of the *'āmil* was known as *ṭappadār*. He was subordinate to the *'āmil* and *amīn*. There could be more than one *ṭappadār* working under an *'āmil* and *amīn*. The *ṭappadār* was supposed to watch over every village and to ensure tax realization by whatever method (*ẓabṭ*, *baṭā'ī* or *kut*), as fixed in the *paṭṭa*.[68] His main duty consisted of obtaining the state's share of the grain (*laṭā*)[69] in the villages whenever the crops

67. *Arhsatta potdār pargana* Gaonri *VS* 1722/1665. *Arhsatta potdār, pargana* Amber *VS* 1720/1663.
68. *Chiṭṭhīs* dated *VS* 1782/1725 ; *VS* 1787/1730 ; *VS* 1813/1756.
69. *Chiṭṭhī Miti Mah vadi* 7, *VS* 1722/1665.
The system of *laṭā* (*lāto* or *latai*) was very popular in Eastern Rajasthan. In cases the cultivator complained of unfair increase or over-estimation in assessment in a given field, he had the choice to have the crop cut and weighed. Thus the *laṭā* seems to be an estimate of the sheafs. The sheafs were shared between the state and the

ripened. He was required to bring together the *paṭel* and the *ra'iyat* and to arrange far *taqāvī* (*tagai*) loans from the moneylender (*bohra*) so that no arable land in any village was left fallow.[70] Apart for *lataī*, he could also assess the land by measurement with the prescribed *dori* (measuring rope).[71]

Like the *potdār*, the *ṭappadār* was also a *mahīnadār*. The rate of his salary is given as Rs. 8/-per month in 1664.[72] In addition to his salary he was entitled to 8 *takas* daily when he was engaged with the *latāī* of a village.[73] He was usually formally instructed by the *'amīl* not to extort from the *ra'iyati* anything extra which was not sanctioned.

Shahna

The *shahna* was perhaps the lowest official—if he can be so called—in the hierachy of the revenue administration. He was appointed in each village by the *rāj* on the recommendation of the *'āmil*. The appointment of the *shahnā* was usually made for one or both seasons of the year.[74] It is also evident from the

peasants. It is possible that *lank batāī* which was an indigenous system might have been corrupted into *lāta* in Rajasthan. The word *latā* or *lātha* is used in the sense of stalks of grains.

70. *Chiṭṭhī Miti Asarh Sudi* 9, *VS* 1804/1747.
71. *Chiṭṭhī* dated *Miti Kati vadi* 4, *VS* 1776/1719. Anūp Chand and Devi Das (*amīn* and *'āmil* respectively) inform *dīwān* Kishor Das and *dīwān* Tara Chand "that the *tappadār* had been sent for measurement (*dorī map*). But the *ra'iyatī* of village Mairgi, *pargana* Bhangarh, called the *patel* and asked him to press the *tappadār* to leave immediately. So the *tappadār* came back." Usually, the *dorī* was of 75 *haths*. A *patel* lodged a complaint in the office of the *dīwān* that his land in *in'ām* had been measured by mistake. In response, the *dīwān* instructed Shri Jai Chand and Shri Har Narain (*amīn* and *'āmil* respectively) that in case a mistake has been committed on behalf of the *tappadār*, it should be rectified.
72. *Arhsatta pargana* Chatsu, *VS* 1721/1664 ; Shyam Tiwari etc. (8 in number), *tappadārs* under Bansidhar, *amīn pargana* Chatsu were paid Rs 64/- and Bhoodhar etc. (8 in number under *Sah* Ajit Das *kārkūn*), *pargana* Chatsu were paid Rs. 64/- per month.
73. *Chiṭṭhī* dated *Miti Vadi* 7, *VS* 1722/1665.
74. *Chiṭṭhī Miti Kartik Vadi* 8 *VS* 1808/1751. *Shahna* Mansa Ram Rikhav Das writes to *dīwān* that the village 'Khoh' *tappa* Ramgarh, *pargana* Sawai Jaipur, where he is working as *shahna* has now been assigned

records that the number of *shahnas* appointed in a *khalīṣa* village was in proportion to the area under *ẓabṭ* and the revenue claimed in kind under *baṭā'ī*.[75] There could be more than one *shahna* in a village. Sometimes, appointment was claimed on a hereditary basis.[76]

The main duty of a *shahna* was to keep watch on the crops when they were ready,[77] or when they were taken to *kotha* (where the grain was stored for the purpose of sale), or to the *ganj* (grain market where it was brought for sale). The person in charge of the *ganj* was called *shahna-i-ganj*.[78] The *shahna* was also required to assist the other officials in the *teḥṣīl* (collection) of revenue.[79]

The *shahna* was entitled to a charge called *shahnagī* from the village. This amounted in all to 15 *mans* of grain for each *shahna*. If there were four *shahnas* in a village, (the area of each *shahna* was specified) then 20 *mans* of grain were to be taken from the village.[80] The excess of the tax (over and above the amount assigned to the *shahnas*) was appropriated by the state under the heading of *ghughri shahna*.[81] A *shahna* who himself cultivated the land was exempted from paying the *ghughri shahna*.[82]

in *jāgīr*. Therefore it is requested that he may be appointed *shahna* of some other *Khāliṣa* village for the whole year (*syalu* and *unahlu*).

75. *Chiṭṭhī Mitī Kartik Vadi* 7, *VS* 1788/1731. *Chiṭṭhī Asarh Vadi* 13, *VS* 1751/1704.
76. *Sanad Miti Bhadva Vadi VS* 1806/1749. Sri Ram Navgi complained to *dīwān* Rai Narain Das : "My father named Daya Ram Navgi was having *shahnagī* in a village of *pargana* Sawai Jaipur. Now he has expired. But the *'āmil* is demanding the *sanad* from me. Kindly instruct the *'āmil* not to demand it".
77. *Chiṭṭhī Miti Chait Vadi* 9, *VS* 1722/1665. The *'āmil* informs the *dīwān* that the *shahna* was ill-treated by the *pāltī* (cultivator) : "three horses were grazing in the field. The *shahna*, appointed there, led them out of the village. The *shahna* just asked a *gunwār* (cultivator) who was irrigating from a well near that field, why the horses were not led out from the field by him, & c."
78. *Chiṭṭhī Miti Kartik vadi* 7, *VS* 1788/1731. *Chiṭṭhī VS* 1805/1748.
79. *Chiṭṭhī VS* 1788/1731.
80. S.P. Gupta, 'The System of Rural Taxation', *PIHC*, Muzaffarpur, 1972, pp. 283-89.
81. *Ibid.*
82. *Arhsatta pargana* Malarna, *VS* 1787/1730.

The Hereditary Village Officials

The villages in Eastern Rajasthan duly contained the pair of village officials who are such a universal feature of Indian village organisation : the headmen (*muqaddam* or *paṭel*) and the accountant (*paṭwārī*). Our records shed considerable light on their positions and functions ; and even if these may not reveal any special regional peculiarities, these are important for showing how the agrarian administration functioned at its lowest level.

Paṭel

The higher levels of the village structure of Eastern Rajasthan comprised the *paṭel, patwārī', shahna, mahājan (bohra, sāhūkār)*, and the *panch* (arbitrators) who were the principal personages involved in the collection of tax from the *pālti* (cultivator). In the second half of the 17th century we also begin to find the word *muqaddam* used for the chief village official, and so synonymous with the word *paṭel*. But the *paṭel* was the most usual designation.[82] The *paṭel* was commonly entitled to the so-called *muqaddamī* rights.[83] He is also referred to as *mukhia*.[84]

The *paṭel* was normally a peasant (*ra'īyatī*) and sometimes a petty *zamīndār*.[85] On the one hand, he was the natural leader of the villagers, and on the other he was the official headman (*mukhia*). He represented the *ra'īyatī* and forwarded their complaints to the *pargana* officials.

82. *Chiṭṭhī Miti Baisakh Vadi* 6, *VS* 1722/1665. In this letter we have a reference to the *muqaddam* and the *paṭel* simultaneously. Also see *Chiṭṭhī* dated *Miti Phalgun Sudi* 9, *VS* 1722/1665.
83. *Chiṭṭhī VS* 1724/1667.
84. *Chiṭṭhī VS* 1723/1666.
85. *Chiṭṭhī VS* 1774/1717 ; *Chiṭṭhī Baisakh Vadi* 6, *VS* 1724/1667; *Chiṭṭhī, VS* 1797/1740 : The *paṭel* complains to the *dīwān* that he has *zamīndārī* rights in four villages in *pargana* Khohri, namely, Vairadh, Jai Singhpur, Pali ka Prannathpur and Ram Singhpura, but the Meos are creating troubles in these villages. "The *dīwān* instructed the *'āmil* to oust the Meos from those villages and take a *muchalka* from the *paṭel*. See also *Chiṭṭhī Miti Sawan Sudi* 1, VS 1723/1666. *Chītthī VS*. 1798/1741.

The office of the *paṭel* was usually hereditary, but the state reserved the right of appointing any person it chose. In addition to a person laying hereditary claim to the office, the appointment could be given to a man who was responsible for the foundation of a new village and was ready to give *peshkash* to the state.[86] The office might also be conferred on a person who could settle a ruined village by his personal efforts.[87] Sometimes the rights of *paṭelī* could be transferred to some other person due to the long absence of the previous incumbent.[88]

In Eastern Rajasthan the *paṭels* often belonged to the caste of Minas and Ahirs, but they were also, though rarely, non-peasants, e.g. *banias* and Brahmans.[89] Jāts are also often found to occupy the office of *paṭel* in this area.[90] There could be one

86. *Chiṭṭhī* dated *VS* 1798/1741, being the *parwāna* of *paṭelī* issued to a man who founded a new village and was prepared to give *peshkash*. *Chiṭṭhī Miti Bhadva Vadi* 10, *VS* 1806/1749. It gives the details of the foundation of the new village, Sawai Ishwari Singh situated within village Lalūras, *tappa* Ramgarh, *pargana* Sawai Jaipur, dated *Unahlu VS* 1803/1747 ; the area of the village was 1450.2.0 (*bīghas*). An amount of Rs. 501/- was fixed as *peshkash*. Five *patels* were appointed : 1. Har Ram Mina, 2. Rajū Mina, 3. Hathū Mina, 4. Dalū Mina, and 5. Thakursi. Hathila *mahājan* became the *patwārī* of the newly founded village. Also see *Chitthī VS* 1723/1666.
87. *Chiṭṭhī Mitī Sawan vadi* 13, *VS* 1776/1719.
88. *Chiṭṭhī Pos Sudi* 12, *VS* 1813/1756. Nanhu Bhagwan Mina was the *paṭel* of village Sankat, *ṭappa ḥavelī, pargana* Bahatri. A Brahman performed the functions of *paṭel* in his place because of his long absence. But the Mina claimed his right when he came back. The Brahman refused to restore the right. The matter was reported to the *dīwān* who instructed the *pargana* officials to see that the *patelī* be restored to the Mina because it had come to him from his forefathers' times. Also see *Chiṭṭhī vadi* 5, *VS* 1783/1726 ; *Chiṭṭhī Miti Bhadva vadi* 10, *VS* 1806/1749.
89. *Chiṭṭhī Pos Sudi* 12, *VS* 1813/1656. Nanhu s/o Bhagwan *Mina* complained that he was assigned *patalaī* (the office of *patel*) of village Sankat, *tappa ḥavelī, pargana* Bahatri (Baswa). Now a Brahman had taken over this office by force.
90. *Raqbo pargano* Mauzabad *VS* 1698/1641. See *Khasras* of different villages. Also see *Yāddāsht asāmī hal bail* of different *parganas*. Cf. S.P. Gupta, 'Agriculture and Agrarian Society, Mid 17th Century Information from Records of *Pargana* Mauzabad in Rajasthan; Table I, *PIHC*, Bodh Gaya, 1981, pp. 370-376.

or more than one *paṭel* in a village. When their number tended to increase the areas of their jurisdiction were separately specified. The *chaudhurī* sometimes also held the office of *paṭel*.[91]

The *paṭel* was required to help the *pargana* officials in the work of revenue assessment and collection. Sometimes he actively participated in the work of measurement along with the *ṭappadār*.[92]

The *paṭel* and the *patwārī* together undertook collection of taxes such as *malba*,[93] *nalbat*,[94] *virad*,[95] from the villagers. The *paṭel* was exhorted to help in bringing untilled land under cultivation and was authorised to recommend grant of *paṭṭas* of such lands to cultivators.

The *paṭel* seems to have played an important role in the farming system (*ijāra*) in Eastern Rajasthan. Generally, he himself took the lands on *ijāra*, either from the *rāj* or from the big assignees. On the other hand, it was his duty to inform the *rāj* about any illegal actions and the excesses of the *jāgīrdār* in a village.[96]

A great deal of responsibility devolved upon the *paṭel* in times of famine. Together with the *chaudhurī* and the *qānūngo* it was his duty to reinhabit the village, to bring to *pāltīs* (cultivators) back to their fields, and to make provisions for seeds, bullocks and ploughs by loans either from the local *bohra* or from the *rāj*.[97] He was required to recommend grants of agricultural loans (*takai=taqāvi*) by the state in times of distress. He was also to exercise vigilance in the supply of water for irrigation purposes (*bandh birar*).

The *paṭel* was supposed to be an essential member of the village 'corporation'. He took part, by the side of the *chaudhurī* and *qānūngo*, in settling disputes over land, and was always present when *chakbandī* or land demarcation was undertaken.[98]

91. *Chiṭṭhī Miti Pos Sudi* 4, *VS* 1723/1666. Cf Dilbagh Singh's article on the 'Position of the *Patel* in Eastern Rajputana during the 18th Century', *PIHC*, Jabalpur, 1970, pp. 360-366.
92. *Chiṭṭhī VS* 1801/1744.
93. S.P. Gupta, 'The System of Rural Taxation' op. cit.
94. *Ibid.*
95. A tax imposed on the *bunds* on water channels.
96. *Chiṭṭhī Miti Sawan vadi* 2, *VS* 1804/1747. *Chitthī VS* 1804/1747.
97. *Chiṭṭhi Baisakh Sudi* 3, *VS* 1722/1665.
98. *Chiṭṭhī Miti Asarh vadi* 6, *VS* 1775/1718.

In case new units of measurement were enforced, such as ***nava hal*** (new plough) instead of *purana hal* (old plough), he was asked to rigorously conform to the new unit.[99] Both the *paṭel* and the *paṭwārī* were consulted by the *'āmil* and the *amīn* at the time of fixing the amount of *peshkash* and *agotri* (state share in *batā'ī* over and above *māl*).[100]

The papers of *bāqi* and *ijāra* were duly signed by the *qānūngo* and the *paṭel*. Sometimes, the *paṭel* was required to prepare the *jama'bandī* and *k͟hasra* papers of the village with the help of the *patwārī*.[101]

The *paṭel* was not paid directly from the treasury of the *rāj*, but he was entitled to some customary sources of income such as *in'ām muqaddamī*,[102] *bisondh* (one-twentieth part),[103] *dagli*[104]) Rs. 1/4/- at the time of the foundation of a new village), *paṭelaī kā burotro*,[105] *siropa* or *siropai* (complete dress of honour received from the Durbar).[106] He was given a *parwāna* for *nānkār*,[107] an allowance of cash or land in lieu of his services. He also received perquisites such as the *paṭelī kā khera kā dol*, a customary amount paid to the *paṭel* on the occasion of the departure of marriage processions from the village.[108] Along with the *patwārī* and the *shāhna* he used to receive a coconut at the time of *baṭā'ī* (sharing of the harvest).[109]

Apart from these and the usual customary *mujrāī*' (deduction from revenue collected), the *patel's* holding was assessed at usually a lower rate in comparison with that of the *ra'iyatī*. Nor

99. *Chiṭṭhī* of *dīwān* to *'āmil* and *amīn* of *pargana* Chatsu *VS* 1723/1666.
100. *Chiṭṭhīs* of different *parganas* of different years.
101. *K͟hasra VS* 1853/1796 of village Madsudanpura, *pargana* Chatsu. Dilbagh Singh, 'Position of the *paṭel*' op. cit.
102. *Arhsatta pargana* Malarna *VS* 1787/1730.
103. See *arhsatta pargana* Chatsu *VS* 1787/1730 for the amount of *in'ām bisondh* received from each village.
104. *Chṭṭthī VS* 1793/1736.
105. *Chiṭṭhī Baisakh Sudi VS* 1785/1728.
106. *Arhsatta* Chatsu *VS* 1721/1664.
107. Dīp Chand and other *paṭels* received the *parwāna* for *nānkār* from the Mahārājkumar. *Arhsatta* Malarna *VS* 1787/1730. There is also evidence of a *mahājan* being granted the revenue of a village in *nānkār*, under the heading of *punya udik* : *Arhsatta* Chatsu *VS* 1787/1730.
108. Dilbagh Singh, 'Position of *Patel*', op. cit.
109. *Chiṭṭhī VS* 1788/1731.

were any *jihāt* or additional dues and perquisites taken from him (Chapter IV).[110]

Patwārī

The *patwārī* was the village accountant who prepared the *khasra* papers[111] as well as accounts of revenue collections. His papers were all in the local language. In the *khasra* papers he was required to record the name of the *asāmī* (cultivator), the area of his holdings (*'arẓ* and *ṭūl*), cultivable area, names of the crops sown in *syalu* (*khārīf*) and *unahlu* (*rabī'*) and the share of *ra'iyatī* and the *dīwān* (state) in each category of land.[112] He was required to assist the *paṭel* and *pargana* officials in the work of revenue collection, measurement of lands and assessment of taxes upon *ra'iyatī*, such as *malba*.[113]

The *patwārī* was required to assist the *paṭel* in the assessment of *malba* and was entitled to a nominal share in it. There were definite instructions from the *rāj* that the assessment of *malba* was to be made in the presence of the *jāgīrdār*, the *paṭel* and the *patwārī*.[114]

110. In addition to the evidence given for such concessional rates in Chapter IV, we reproduce here a table of comparative rates from *Arhsatta pargana* Banawar *VS* 1770/1713 :

Crop (*kharīf*)	*Rate of assessment* *Patel*	*(in rupees per bīgha)* *Ra'iyatī*
Cotton (van)	2. 4.0	2. 8.0
Paddy (dhan)	4.12.0	5. 4.0
Barai	3. 0.0	3. 8.0
Varh (sugarcane)	4.12.0	5. 8.0
Jowar	5. 8.0	6. 0.0
Moth	3. 4.0	3.12.0

111. *Khasra* dated *VS* 1853/1796 ; *sakh syalu* of village 'Madusudanpura, *pargana* Chatsu. It is signed by the *paṭel* and *paṭwārī* named Karan Chand *bania*.
112. *Ibid*. Also see *khasra mauẓa* 'Tuteo *VS* 1768/1711.
113. *Chiṭṭhī VS* 1804/1747.
114. *Chiṭṭhī VS* 1804/1747.

It is not clear from the records whether the *patwārī* was paid by the peasants or by the state. It appears that he enjoyed certain rights and privileges such as the *dastūr-patwārā* levied upon the cultivator ; he also took the *dastūr* from the merchants for weighing the grain, this being part of his duty.[115] The amount realised under the tax *patwārā* was shared by the *patwārī* and the state equally. The proportion of this share was indicated in the *paṭṭa*.[116] Usually the *patwārā* was levied on the peasants (*ra'īyatī*) and merchants at different rates. Sometimes the right to collect the *patwārā* was farmed out by the *patwārī*.[117]

The *patwārī* was required to give *peshkash* (*peshkash patwārā*) at the time of his appointment and also at the time of the foundation of a new village.[118] In *ujār* (deserted) villages no *patwārī* was appointed ; it was simply cultivated by the *pā'is* or 'outsiders'. Like the *paṭel*, the *paṭwārī* used to receive a robe of honour at the time of appointment.[119]

Like the *paṭel*, the *paṭwārī* was also assessed at a lower rate upon his lands. On the *bichoti* (grains sold) too a concessional duty *ekotra*, Re. -/5/2% (upon the sale of one hundred *mans*) was charged from him, while from the merchants the levy amounted to -/10/2%.[120]

115. He used to get *takā* 6 per *man* as *tulai*.
116. *Chiṭṭhī* dated *Miti Chait Sudi* 3, *VS* 1702/1645. A complaint was lodged by the *ra'īyatī* against the *patwārī* for his excessive claims to *patwārī*. The *patwārā* was instructed to charge it as stipulated in the *paṭṭa*.
117. *Ibid*., *Chiṭṭhī VS* 1722/1665.
118. *Chiṭṭhī* dated *VS* 1743/1686 for the appointment of *patwārī*, *pargana* Geejgarh. *Parwana* dated 18 *Rabi'ul Awwal* Year 1098 for the appointment of *patwārī* with amount fixed upon him as *peshkash*. *Chiṭṭhī Miti Bhadva Vadi* 10, *VS* 1806/1749 : the appointment of Hathila *mahājan* as *patwārī* of Sawai Ishwari Singh, a newly founded village in *ṭappa* Ramgarh, *pargana* Sawai Jaipur and the *peshaksh* fixed upon him.
119. *Arhsatta* Chatsu *VS* 1721/1664. *Arhsatta* Chatsu *VS* 1787/1730,
120. At the time of the foundation of village Mudaji etc., Ganga Ram *paṭel* etc. (total number 4) were given *siropa* as Rs. 12.15.0 as per details given below :

Hira Yogi	Raju Hīra *paṭel* Mina
	pao (*dhotī*) 1
Rs. 8.0.0	Rs. 1.13.0
Lavakas, Mina pao 1	Paima, Mina *paṭel*

In examining our evidence for the structure of revenue administration in Eastern Rajasthan, we are impressed forcibly by the wholesale application of the standard Mughal system in so far as we can reconstruct the latter from the official manuals of that Empire. It almost seems as if the bureaucrats of the Kachhwaha principality had not only inherited that system (which must, of course, have been the case, where the Jaipur house took over *maḥals* which were previously in the *jāgīr* of other Mughal nobles), but they also aimed at continuing the same routine and regulations. The changes, if any, seem to be not even of detail, but only of context. Often enough, the Jaipur Court had to act both as a *jāgīrdār* and as a successor to the Mughal Imperial Court. The two roles naturally fused here, though these remained distinct in the main period of the Mughal Empire. But the contrast would disappear if we remember that even in that Empire, these roles were united into one in the administration of the emperor's *khāliṣa*, which was essentially the emperor's own *jāgīr*. In the next chapter, we would meet the Jaipur ruler in that position, where he had to act more in the manner of the Mughal sovereign on a smaller scale—that is when he himself had to assign and control *jāgīrs* within his principality.

Rs. 1.9.0 1.9.0
Patwārī etc. pao white 2 for Rs. 1.9.0
Akhari Ram *patwārī* Hirabalai pao 1
0.12.2 0.12.2

121. *Chiṭṭhī VS* 1735/1678.

APPENDIX

I. *List of Dīwāns* (c. 1650-1750)

Name	*Year (AD)*
Narain Das	—
Gokal Das (*dīwān desh* or *dīwāngirī desh*)	—
Raja Ram and Tej Singh	—
Dīwān Shri Kalyan Das	1664
Dīwān Gaj Singh and Maha Singh	1665
Dīwān Shri Kalyan Das	1667
Dīwān Raja Ram and Tej Singh	1667
Dīwān Kishor Das and Vijai Ram	1692
Dīwān Kishor Das Laxmi Das	1705-1706
Sah Fateh Chand	1708
Dīwān Bhikhari Das and Fateh Chand	1709-1713
Dīwān Sri Chand	1713 (for a short period)
Dīwān Rai Narain Das Kirpa Ram	1716
Dīwān Kishore Das and Tara Chand	1717
Dīwān Kishore Das and Tara Chand	1719
Dīwān Tara Chand	1721
Dīwān Rai Narain Das Kirpa Ram	1725
Dīwān Rai Narain Das Kirpa Ram	1728
Dīwān Rai Narain Das Kirpa Ram	1731
Dīwān Rai Narain Das Fateh Chand (Ram)	1731 (for some time)
Dīwān Vidhyadhar	1731 (for some time)
Dīwān Rai Narain Das Vidhyadhar	1737-38
Dīwān Vidhyadhar and Fateh Ram	1739-41
Dīwān Rai Narain Das Bakht Ram	1742
Dīwān Murli Dhar	1744
Dīwān Rai Narain Das Fateh Ram	1745
Dīwān Rai Narain Das Har Govind	1748
Dīwān Vidhyadhar	1749
Dīwān Har Govind Kanhi Ram	1750-51

II.(a) List of 'Āmil, Amīn and Kārkuns
(c. 1650-1750)
Pargana Amber (Sawāī' Jaipur)

1	2	3	4	5
Year	*'Āmil*	*Amīn*	*Kārkun*	*Period of appointment*[122]
1677	—	Alauddin	—	1 Rajab 1088 to the last month of the year 1089.
1688	—	Ajit Das	Kalū Ram	—
1689	—	Prohit Harnam	Ajit Das	1 *Ẓiqadah*, 1100 to *Ẓiqadah* 1102.
1712	*Sah* Nain Sukh[123] *Sah* Sri Chand *Sah* Nanig Ram	Bhayya Kishor Das	—	*Miti Bhadva Sudi* 3, *VS* 1769 to *Miti Bhadva Sudi* 2, *VS* 1770.

122. After the general survey of these Tables of different *parganas* in different years, it is interesting to note that after the death of Aurangzeb the *rāj* of Amber introduced the *Vikrama Samwat* in the records along with the *Hijri era*. But soon after the death of Bahadur Shah, it removed the *Hijri era* and retained only the *Vikrama Samvat*.

123. The details of their tenure are as follows :

Sah Nain Sukh	Bhayya Kishor Das	*Miti Bhadva Sudi* 3, *VS* 1769 to *Katik Sudi* 2 (2 months).
Sah Sri Chand	,,	*Miti Kartik Sudi* 3, *VS* 1769/1712 to *Miti Asadh Vadi* 6 *VS* 1770/1713 (7 months 19 days).
Sah Nanig Ram	,,	*Miti Asadh Vadi* 7, *VS* 1770/1713 to *Bhadon Sudi* 2, 1770/1713.

Arhsatta mujmili pargana Amber *VS* 1769/1712. Also see *arhsatta mujmili pargana* Amber *VS* 1772/1715.

1715	Dalal Chand Seva Ram Lal Chand	Nanig Ram	—	*Miti Bhadva Sudi* 3, *VS* 1772 to *Miti Bhadva Sudi* 2, *VS* 1773.
1716	*Sah* Sewa Ram	—	—	*Miti Bhadva Sudi* 3, *VS* 1773 to *Miti Bhadva Sudi* 2, *VS* 1774.
1718	*Sah* Sewa Ram *Sah* Mohan Das	—	—	*Miti Bhadva Sudi* 3, *VS* 1775 to *Miti Bhadva Sudi* 2, 1776.
1720	*Sah* Mohan Das *Sah* Sahib Ram	—	—	*Miti Bhadva Sudi* 3, *VS* 1777 to *Miti Bhadva Sudi* 2, *VS* 1978.
1723	*Sah* Sahib Ram	—	—	*Miti Bhadva Sudi* 3, *VS* 1780 to *Miti Bhadva Sudi* 2, *VS* 1781.
1724	*Sah* Sahib Ram	—	—	*Miti Bhadva Sudi* 3, *VS* 1781 to *Miti Bhadva Sudi* 2, *VS* 1782.
1725	*Sah* Sahib Ram	—	—	*Miti Bhadva Sudi* 3, *VS* 1782 to *Miti Bhadva Sudi* 2, *VS* 1783.
1728	—	Vidhya Dhar	—	*Miti Bhadva Sudi* 3, Monday, *VS* 1785 to *Miti Bhadva Sudi* 1 & 2, Thursday 1786.
1729	Khushali Ram	Vidhya Dhar	—	*Miti Bhadva Sudi* 3, *VS* 1786 to *Bhadva Sudi* 2, *VS* 1787.
1733	Khushali Ram	Vidhya Dhar	—	*Miti Bhadva Sudi* 3, *VS* 1790 to *Miti Phalgun Sudi* 12, *VS* 1790.

1	2	3	4	5
1734	*Sah* Nanig Ram	Vyas Achal Ram (Abachal Ram	—	*Miti Bhadva Sudi* 3, *VS* 1791 to *Bhadva Sudi* 2, *VS* 1792.
1736	*Sah* Nanig Ram	*Sah* Nanig Ram	—	*Miti Bhadva Sudi* 3, *VS* 1793 to *Miti Bhadva Sudi* 2, *VS* 1794.
1737	*Sah* Nanig Ram	Sah Nanig Ram	—	*Miti Bhadva Sudi* 3, *VS* 1794 to *Miti Bhadva Sudi* 2, *VS* 1795.
1741	*Sah* Shri Mansa Ram	Raja Ram	—	—
1744	*Sah* Har Narain	Prohit Jai Chand	—	*Miti Bhadva Sudi* 3, *VS* 1801 to *Miti Bhadva Sudi* 2, *VS* 1802
1747	*Sah* Har Narain	—	—	—
1748	*Sah* Har Narain	—	—	—
1749	*Sah* Har Narain	—	—	—
			Pargana Banatri	
1665	—	Jodh Raj	Girdhar Das	—
1685	—	Syo Ram	Syo Dutt	1 *Safar* 1096 to 16 *Ẓilhij* 1097.
1686	—	Syo Ram	Man Roop	1 *Shawwāl* 1907 (H.) to *Asarh Shawwāl*, 1098 (H).
1688	—	Syo Ram Das	Man Roop Ram Chand	—

1689	(Lakmi) Lakshmi Das	Syo Ram Das	—	1 *Ẓiqadah*, 1100 to the last month 1101 (H)
1692	Lakshmi Das	Bhayya Shyam Ram Das	—	—
1696 K.[1]	Hathi Ram	Ram Das	—	1 *Safar* 1108 to 18th of *Ramẓān* 1108,
R.[2]	Ch. Kushal Singh Bhopati Ram	*Sah* Kunwar Pal	—	19 *Ramẓān* to 10 *Muḥarram*
1697	Bhopati Ram	Kunwar Pal	—	1118 to 11th *Shawwāl* 1118
1706	Chatarbhuj Ram Ji	Purohit Harikishan	—	*Bhadva Sudi* 3 to *Miti Bhadva Sudi* 2, 12 *Sawan* 1118 to *Awali* 1119.
1708	*Sah* Anand Ram	*Sah* Ganga Rom	—	*Miti Asoj Vadi* 6, *VS* 1765 *Saal* Year 1120 to 29 of last month year, *Miti Bhadva Sudi* 2.
1710	Har Gobind	*Sah* Anand Ram	—	1 *Rajab* year 1122 *Miti Bhadva Sudi* 3, Tuesday *VS* 1767, *Saal* 1118 to 29 *Rajab* 1123, *Bhadva Sudi* 1 and 2.
1711	Har Govind	Anand Ram	—	*VS* 1768 *Saal* 1119 from 17 *Ẓilhij* 1123 *Miti Bhadva Vadi* 4, Tuesday, to *Asarh* dated 29 *Rajab* 1124, *Miti Bhadva Sudi* 2.

1. K=*K̲h̲arīf*
2. R=*Rabī'*

1	2	3	4	5
1713	*Sah* Har Govind	*Sah* Anand Ram & Daulat Ram	—	*Bhadva Sudi* 3, *VS* 1770 to *Miti Phalgun Sudi* 2, *VS* 1770.
1716	*Sah* Ram Rai	Daulat Rom	—	*Miti Bhadva Sudi* 3, *VS* 1773 to *Miti Bhadva Sudi* 3 *VS* 1774.
1717	*Sah* Ram Rai Maya Ram Bodha	Bijay Ram Sawara	—	*Miti Bhadva Sudi* 3, 1774 to *Miti Bhadva Sudi* 2, *VS* 1775.
1718	Maya Ram Godha	Bijay Ram Sawara	—	*Miti Bhadva Sudi* 3, *VS* 1779 to *Miti Bhadva Sudi* 2, *VS* 1776.
1720 K.	Maya Ram *Sah* Lunkaran	Purohit Malūk Chand	—	*Miti Bhadva Sudi* 3, Thursday, *VS* 1777 to *Miti Chait Vadi* 5, Tuesday & from *Miti Phalgun*
R.	*Sah* Lunkaran *Sah* Hans Raj	Purohit Malūk Chand	—	*Sudi* 3, Saturday to *Miti Bhadva Sudi* 1 & 2 Sunday.
1721	*Sah* Hans Raj *Sah* Dalu Ram	Malook Chand	—	*Miti Bhadva Sudi* 3, Monday *VS* 1778 to *Miti Chait Vadi* 2, Wednesday *VS* 1778.
1724	*Sah* Dulli Chand	Purohit Raja Ram	—	*Miti Bhadwa Sudi* 3, Monday *VS* 1781 to *Miti Bhadva Sudi* 2 & 3, Saturday *VS* 1782.
1725	Dulli Chand	Purohit Raja Ram	—	*Miti Bhadva Sudi* 2, *VS* 1782

1740	Chhokh Chand Ajit Das	Purohit Anand Ram	—	*Miti Phalgun Sudi* 13, Monday *VS* 1797 to *Bhadva Sudi* 2, *VS* 1797.

Pargana Malarna

1690 K. & R.	—	Sanghi Shiv Datta (Syo Dutt)	*Sah* Pirag Das	1 *Ẓilhij* 1101 to *Ziqa'd*, 1102 —
1699 R.	—	Bhikari Das Nathu Ram	Sri Dhar	1 *Ramẓān Avli* year 1111 to last month *Sha'bān* year 1111.
1771 K. & R. 1712	*Sah* Dala Ram Banarsi Das Dala Ram (*Bhadva Sudi* 3 to *Miti Asoj Sudi* 7) for 36 days Banarsi Das Surati Ram Also *gumashta* Nusrat Yar Khan (for 20 days)	*Sah* Gulab Chand Dala Ram Narain Das	— —	1 *Sha'bān* (1113 to *Rajab** *Miti* Bhadva Sudi 2, VS 1768* 1124 whole year

1	2	3	4	5
1713 K.	Surati Ram Manohar Das	Narain Das	—	Whole year
R.	Brij Lal	*Sah* Ahmar Chand	—	
1714 K.	Ranvir Amar Chand	Bakht Ram Narain Das	—	*Miti Bhadva Sudi* 3, 1771 Tuesday to *Miti Bhadva Sudi* 2, 1772. Saturday.
R.	Malūk Chand Dosi (may be Joshi) Bandi Das	Bakht Ram		
1715 K.	Manohar Das	Roop Chand	—	*Miti Bhadva* Sudi 2 (six months)
R.	Manohar Das	Roop Chand	—	*Miti Phalgun Sudi* 3 to *Bhadva Sudi* 2. *Vs*. 1772.
1716 K.	Manohar Das	Roop Chand	—	*Miti Bhdava Sudi* 3, to *Phalgun Sudi* 2, *Vs* 1773.
R.	Nanig Ram	Roop Chand Narain Das	—	*Miti Phalgun Sudi* 3, to *Miti Bhadva Sudi* 2
1717 K.	Mozi Ram	—	—	*Miti Bhadva Sudi* 3, to *Miti Phalgun Sudi* 2, Friday
R.	Bhayya Kalu Ram	Mozi Ram	—	*Miti Baisakh Sudi* 12, Thursday to *Miti Bhadva Sudi* 2, *VS* 1717

1718 K.	Bhayya Kalu Ram	Mozi Ram	—	
R.	Bhayya Kalu Ram	*Sah* Narain Das	—	*Miti Bhadva Sudi* 3, Monday to *Miti Bhadva Sudi* 2, *VS* 1775.
1719 K.	Bhayya Kalu Ram	*Sah* Shambhu Ram	—	*Miti Ehadva Sudi* 3, *VS* 1776 to *Phalgun Sudi* 2, *VS* 1776
R.	Bhayya Kalu Ram	Shambhu Ram	—	*Miti Phalgun Sudi* 3 to *Bhadva Sudi* 2.
1720 K.	Bhayya Kalu Ram	Bakhat Mal	—	*Miti Bhadva Sudi* 3 to *Phalgun Sudi* 2, *VS* 1777.
R.	Narain Das Girdhar Das	Bakhat Mal	—	*Miti Phalgun Sudi* 3, *VS* 1777 to *Bhadva Sudi* 2, *VS* 1777.
1721 K.	Girdhar Das	Bakhat Mal	—	—
1722 K. & R.	Kesri Singh	Akhey Ram	—	*Miti Bhadva Sudi* 3, *VS* 1779 to *Miti Bhadva Sudi* 2, *VS* 1780.
1723 K. & R.	Kesri Singh	Akhey Ram	—	*Miti Bhadva Sudi* 3, 1780 to *Miti Bhadava Sudi* 2, *VS* 1781
1725	Fateh Ram	Sah Man Ram	—	*Miti Bhadva Sudi VS* 1783 to *Miti Bhadva Sudi* 2, 1784
1726 K. & R.	*Sah* Fateh Ram	*Sah* Man Ram	—	Whole year
1727	Puran Mal	Bhim Raj	—	Whole year.

1	2	3	4	5
1728	Puran Mal	Bhim Raj	—	*Miti Bhndva Sudi* 3, *VS* 1785 to *Miti Bhadva Sudi* 2, *VS* 1786.
1729	Hans Raj	Puran Mal	—	Whole year.
1730	Hans Raj	Puran Mal	—	Whole year.
1731 K.	Hans Raj	Puran Mal	—	*Miti Bhadva Sudi* 3, *VS* 1788 to *Phalgun Sudi* 2.
1732 K.	Hans Raj	Puran Mal	—	*Miti Bhadva Sudi* 3, *VS* 1789 to *Miti Bhadva Sudi* 2, 1790.
R.	Bhayya Ajay Singh	Akhey Ram	—	Whole year.
1736 K.	Mozi Ram	Chhajju Ram	—	*Miti Bhadva Sudi* 3, *VS* 1793 to *Bhadva Sudi* 2, *VS* 1794.
R.	Mozi Ram	Surat Ram	—	
1737 K.	Mozi Ram	Surat Ram	—	*Miti Bhadva Sudi* 3, *VS* 1794 to *Miti Baisakh Sudi* 2, 1794.
R.	Surat Ram	Sehaj Ram	—	*Baisakh Sudi* 3, *VS* 1794 to *Bhadva Sudi* 2, *VS* 1795.
1738 K.	Surati Ram	Sehaj Ram	—	*Miti Bhadva Sudi* 3, *VS* 1795 to *Chait Vadi* 12, *VS* 1795.
R.	Ch. Surati Ram	*Sah* Anand Ram	—	*Chait Vadi* 13, *VS* 1795 to *Miti Bhadva Sudi* 2, *VS* 1795.

1743 K.	Baks Ram	Pratap Singh	—	*Miti Bhadva Sudi* 3, *VS* 1800 to *Chait Vadi* 9.
	Pratap Singh	Nihal Chand		
R.	Udai Ram	Nihal Chand	—	*Chait Vadi* 10 to *Miti Bhadva Sudi* 2, *VS* 1801.
1744 K.	Udai Ram	Nihal Chand	—	*Miti Bhadva Sudi* 3, *VS* 1801 to *Chait Suai* 10.
	Kripa Ram			
R.	Surti Ram	Nihal Chand	—	*Chait Sudi* 11, *VS* 1801 to *Miti Bhadva Sudi* 2.
1745	Surti Ram	Nihal Chand	—	Whole year.
1748	Shobha Chand	Nihal Chand	—	Whole year.
1749	Shobha Chand	Nihal Chand	—	*Miti Bhadva Sudi VS* 1806 to *Miti Bhadva Sudi* 2, *VS* 1807. (Whole year)
1750 K.	Shobha Chand	—	—	*Miti Bhadva Sudi* 3, *VS* 1807 to *Miti Mangsir Sudi* 15, *VS* 1807.
1750	Seva Ram Nahnu Ram	Ram Chand	—	*Chait Vadi* 13, *VS* 1807 to *Baisakh Vadi* 12, *VS* 1808.
		Pargana Dausa		
1709 K.	Santosh Ram	*Sah* Syo Ram Das	—	1 month *Rajab* year 1121 *Bhadva Sudi* 3, *VS* 1766 the 26 *Muḥarram* year 1122,

1	2	3	4	5
R.	Syo Ram Das			*Chait Vadi* 13 and 27th *Muḥarram* 1122, *Chait Vadi* 14 to 29 *Jama'di-us Ṣānī* year 1122, *Miti Bhadva Sudi* 2, *VS* 1767.
1712 K.	Tek Chand	Nanig Ram	—	*Miti Bhadva Sudi* 3, *VS* 1769 to *Pos Sudi* 4, Friday *VS* 1769 year 1120 and from *Miti Pos Sudi* 5, Saturday, to *Bhadva Sudi* 1 & 2 Monday *VS* 1770.
R.	Kushal Chand	Nanig Ram		
1713	Kushal Chand	Nanig Ram	—	*Miti Bhadva Sudi* 3, *VS*, to *Bhadva Sudi* 2, *VS* 1771.
1715	Fateh Chand	Nanig Ram	—	*Bhadva Sudi* 3, *VS* 1772 to *Bhadva Sudi* 2, *VS* 1773.
	Bhayya Ram	Fateh Chand		
	Fateh Chand	—		
1716	Sirkat Ram in place of Fateh Chand			—
1718 K.	Mīm Raj	Bhayya Kanha Ji	—	—
1736	Syo Ram	Bhayya Kripa Ram	—	—
1738	Bhayya Kirpa Ram	*Sah* Sukh Ram	—	*Miti Bhadva Sudi* 3, 1775 to *Miti Chait Vadi* and from *Miti Chait Sudi*, *VS* 1775 to *Miti Bhadva Sudi* 2, *VS* 1775.

1739	*Sah* Nihal Chand Bhayya Raj Ram	*Sah* Amar Chand Mozi Ram	—	*Phalgun Vadi* 3, *VS* 1796 to *Bhadva Sudi* 2, *VS* 1797.
1741	Govind Ram	Mozi Ram	—	*Miti Bhadva Sudi* 3, Monday *VS* 1798 to *Kartik Vadi* 5, Friday *VS* 1798.
1742	Ch. Malūk Ram *Sah* Deep Chand	*Sah* Deep Chand Ch. Malūk Ram	—	*Bhadva Sudi* 3, *VS* 1779 to *Chait Sudi* 15 and from *Baisakh Vadi* 1, *VS* 1799 to *Bhadva Sudi* 2, *VS* 1800.
1745	Bhayya Hiraman	Santosh Ram	—	—
1746	*Sah* Gumani Ram	Purohit Santosh Ram	—	*Miti Bhadva Sudi* 3, Tuesday, *VS* 1803 to *Miti Jeth Sudi* 8, *VS* 1804.
1747	*Sah* Puran Mal	—	—	*Miti Bhadva Sudi* 3, *VS* 1804 to *Miti Asoj Vadi* 6 and *Miti Asoj Vadi* 7, Saturday *VS* 1804 to *Miti Bhadva Sudi* 2, *VS* 1804.
1750	*Sah* Dalū Ram *Sah* Jiwan Ram	— —	—	—
		Pargana Lalsot		
1687	—	Bhikari Das Bhopati Ram Hem Raj	—	1 month *Ẓiqa'd* A.H. year 1098 to the last of *Shawwāl* year 1099.

1	2	3	4	5
1713 K. & R.	Mansa Ram Bhopati Ram	*Sah* Kishor Das Moti Ram	—	*Miti Bhadva Sudi* 3, *VS* 1770 to *Miti Sawan Vadi* 5, *VS* 1771. (H. 1121).
1714	Ch. Kirpa Ram	Sah Ramanand	—	*Miti Bhadva Sudi* 6, *VS* 1771 to *Miti Bhadva Sudi* 2, *VS* 1771.
1715 K. & R.	Ch. Kirpa Ram Bhopati Ram	Ramanand Ch. Kirpa Ram	—	*Miti Bhadva Sudi* 3, *VS* 1772 to *Miti Jeth Vadi* 13, *VS* 1773 and *Miti Jeth Vadi* 14, to *Miti Bhadva Sudi* 2, *VS* 1774.
1717 K. R.	Bhopati Ram Narain Das	Ch. Kirpa Ram Narain Das Khushal Chand	—	*Miti Bhadva Sudi* 3, 1774 to *Miti Baisakh Sudi* 9, *VS* 1775 *Miti Baisakh Sudi* 10, *VS* 1775 to *Bhadva Sudi* 2, *VS* 1775.
1718	Narain Das Ch. Daulti Ram	Khushal Chand	—	*Miti Bhadva Sudi* 3, Sunday *VS* 1775 to *Miti Sawan Sudi* 11, Thursday, *VS* 1776.
1719	Ch. Daulti Ram	Bhopati Ram Kishor Das	—	*Miti Bhadva Sudi* 2, *VS* 1776 to *Miti Bhadva Sudi* 2, *VS* 1776.
1720	Ch. Sada Ram	Kishor Das *Sah* Ganga Ram	—	*Miti Bhadva Sudi* 3, *VS* 1776 to *Miti Bhadva Sudi* 2, *VS* 1777.
1721 K.	Ch. Sada Ram Sah Sri Chand	*Sah* Ganga Ram Ch. Sada Ram	—	*Miti Bhadva Sudi* 3, *VS* 1778 to *Miti Phalgun Sudi* 12 *VS* 1778.

1721 R.	Tirlok Chand	*Sah* Sri Chand	—	*Phalgun Sudi* 13, to *VS* 1778 to *Asadh Vadi* 9, *VS* 1778.
	Bhayya Daya Ram	Ch. Kishan Ram		
1722	Bhayya Daya Ram	Ch. Kishan Ram	—	*Miti Bhadva Sudi* 3, *VS* 1779 to *Miti Bhadva Sudi* 2, *VS* 1780.
1726	*Sah* Har Ram	Purohit Har Chand	—	—
1730	*Sah* Lal Ram	*Sah* Raja Ram	—	*Miti Bhadva Sudi* 3, *VS* 1787 to *Miti Phalgun Sudi* 2, *VS* 1787.
1731	*Sah* Lal Ram	Nanig Ram	—	*Miti Bhadva Sudi* 3, *VS* 1788 to *Miti Bhadva Sudi* 2, *VS* 1789.
1733	Ajab Ram	Roop Chand *Sah* Than Singh	—	*Miti Bhadva Sudi* 3, *VS* 1790 to *Miti Bhadva Sudi* 2, *VS* 1791.
1734	Ajab Ram	*Sah* Than Singh	—	*Miti Bhadva Sudi* 3, *VS* 1791 to *Miti Bhadva Sudi* 2, *VS* 1792.
	Ram Chand	*Sah* Than Singh		
1737	*Sanghi* Ram Chand	Shobha Chand	—	*Miti Bhadva Sudi* 3, *VS* 1794 to *Bhadva Sudi* 2, *VS* 1795.
1739 K.	Champa Ram	Shobha Chand	—	*Miti Bhadva Sudi* 3, *VS* 1796 to *Miti Bhadva Sudi* 2, *VS* 1797.
R.	Bhakht Ram			
1740 K.	*Sah* Bhakht Ram	Shobha Chand	—	*Miti Bhadva Sudi* 3, *VS* 1797 to *Miti Bhadva Sudi* 2, *VS* 1798.
R.	Ajab Ram			
1741	Ajab Ram	Shobha Chand	—	*Miti Bhadva Sudi* 3, *VS* 1798 to *Miti Sawan Vadi* 7, *VS* 1798.

1	2	3	4	5
1745	*Sah* Shobha Chand	Das Nirbhay Das	—	*Miti Bhadva Sudi* 3, *VS* 1802 to *Miti Bhadva Sudi* 2, *VS* 1803.
1746 K.	*Sah* Shobha Chand	Das Nirbhay Ram	—	*Miti Bhadva Sudi* 3, *VS* 1803 to *Miti Bha va Sudi* 1 & 2, *VS* 1804.
R.	*Sah* Shobha Chand	Sah Jiwan Ram	—	*Miti Bhadva Sudi* 1 & 2, *VS* 1803.
1747	—	Miyan Sahib Khan	—	*Miti Bhadva Sudi* 1 & 2, *VS* 1804. Whole year.
1748	—	Miyan Sahib Khan	—	Whole year
		Pargana Chatsu		
1664	—	Bansi	Ajit Das	18 *Rabi'-ul Awwal* 1075 to *Ramẓān* 1075.
1689	*Sah* Fateh	—	—	—
1694	Angat Ram	Sah Man Ram Nitani	—	1 *Muḥarram* 1106 to last of *Muḥarram* 1107.
1695	*Sah* Har Ram	Rūp Chand	—	—
1708 K. R.	Kunwar Pal	Kushal Singh	—	1, 1120 to 20 *Muḥarram* 1121.
1709 K. +R.	Khushal Chand	Kesri Singh Naroda	—	1 *Rajab* 1121 to 16 *Awwal* 1122.
1710	Maun Sukh	—	—	—

1711	Nain Sukh Megh Raj	*Sah* Hari Singh	—	1 *Sh'ābān* 1123 *Miti Bhadva Sudi* 3, *VS* 1768 to *Asadh Miti Bhadva Sudi* 2, *VS* 1769.
1712	Megh Raj	*Sah* Hari Singh Gulab Chand	—	—
1714	Megh Raj	*Sah* Gulab Chand	—	—
1715	Megh Raj	*Sah* Gulab Chand	—	—
1716	Sahib Ram	Megh Raj *Sah* Amar Chand	—	*Miti Bhadva Sudi* 3, *VS* 1773 to *Miti Bhadva Sudi* 3, *VS* 1773.
1717	Sahib Ram Gulab Chand	*Sah* Amar Chand Kirpa Ram	—	*Miti Bhadva Sudi* 3, *VS* 1774 to *Miti Bhadva Suai* 2, *VS* 1775.
1721	Gulab Chand	Kirpa Ram	—	—
1723	Kanhi Ram	Ch. Kirpa Ram Ch. Bakht Ram	—	—
1724	Kanhi Ram	Ch. Bakht Ram	—	—
1726	Sah Nemi Das, Sanghi Devi Das Sanghi Ajab Singh	Ch. Bakht Ram *Sah* Nanig Ram Mansa Ram	—	*Miti Bhadva Sudi* 3, *VS* 1783 to *Asadh Sudi* 7 *VS* 1783 ; again from *Asadh Sudi* 8 to *Bhadva Sudi* 1 and 2. *VS*. 1784.
1730	Nihal Chand	*Sah* Mukund Das	—	—
1731	*Sah* Ajit Das	*Sah* Mukund Das	—	—
1735	Nihal Chand	Rekh Raj	—	—

1	2	3	4	5
1736	Nihal Chand	Shukh Rekh Raj	—	—
1737	*Sah* Jiv Raj	*Sah* Rekh Raj	—	*Miti Bhadva Sudi* 3, *VS* 1794 to *Miti Bhadva Sudi* 2, *VS* 1795.
1738 K.	*Sah* Jiv Raj	*Sah* Rekh Raj Purohit Ghasi Ram	—	*Miti Bhadva Sudi* 3, Friday, *VS* 1795 to *Asoj Sadi* 13, Tuesday, *VS* 1095.
R.	*Sah* Ganga Ram	Purohit Ghasi Ram *Sah* Rikhuv Das Sanghi Bakht Ram	—	*Miti Asoj Vadi* 14 Wednesday *VS* 1795 to *Pos Vadi* 11 Saturday *VS* 1795, then *Pos Vadi* 12, *VS* 1795 to *Chait Sudi* 5, *VS* 1795 ; *Chait Sudi* 6 to *Sawan Sudi* 15, *VS* 1796.
1739	*Sah* Arat Ram	Fateh Chand	—	—
1741	Surat Ram Sawat Ram	Fateh Chand	—	*Miti Bhadva Sudi* 3, *VS* 1778 to *Miti Asoj Vadi* 1 ; *Miti Asoj Vadi* 2, *VS* 1798 to *Miti Bhadva Sudi* 2, *VS* 1798.
1742	*Sah* Sawai Ram	*Sah* Fateh Chand	—	*Miti Bhadva Sudi* 2, *VS* 1799 to *Miti Bhadva Vadi* 2, *VS* 1800.
1743 R.	Ajab Singh	Kirpa Kam	—	—
1744	Sanghi Ajab Singh	Bhayya Kirpa Ram	—	*Miti Bhadva Sudi* 3, *VS* 1801 to *Jeth Sudi* 12, *VS* 1802.

1745	Sewa Ram *Sah* Ram Chand	Bhayya Kirpa Ram	—	*Bhadva Sudi* 3, *VS* 1802 to *Bhadva Sudi* 1 and 2, *VS* 1803.
1748	Arati Ram	—	—	—
1749	Arati Ram	—	—	—
1750	Arati Ram	—	—	*Miti Bhadva Sudi* 3, *VS* 1807 to *Miti Pos Sudi* 15, *VS* 1807.

II (b)

Pargana Hindaun

VS/A.D.	*'Āmil*	*'Amīn*
1770/1713	Manohar Das Jiv Raj	Bhayya Ram
1771/1714	Mansa Ram Har Govind	*Chaudhurī* Ram, Kirpa Ram, Kishan Chand
1774/1717	Rūp Chand Bhayya Chabil Chand	
1775/1718	*Sah* Daulat Ram	
1777/1720	*Sah* Ajit Singh Sanghi Baks Ram	Anoop Chand
1778/1721	Har Govind Natani Mansa Ram	*Sah* Amar Chand
1779/1722	*Sah* Amar Chand	*Sah* Amar Chand
1780/1723	*Sah* Amar Chand (*mahājan*)	Har Govind Natani
1781/1724	*Sah* Fateh Chand Chabra *Sah* Girdhar Das (*mahājan*) Chabra	
1784/1727	*Sah* Har Ram (*mahājan*)	Bhayya Gulab Chand
1785/1728	*Sah* Dungarsi (*mahājan*)	Bhayya Gulab Chand
1787/1730	*Sah* Dungarsi (*mahājan*)	Purohit Ghasi Ram
1789/1732	Sanghi Devi Das	Purohit Ghasi Ram
1790/1733	*Sah* Jiv Raj	Purohit Ghasi Ram
1796/1739	(*mahājan*) Har Chand	Khushal Chand (*mahājan*)
1803/1746	Ram Kishan (*mahājan*) Natani	Khushal Chand (*mahājan* and Kirat Ram (*mahājan*)
1804/1747	Udai Ram *mahājan*	
1805/1748	Udai Ram *mahājan*	
1811/1754	*Sah* Gulab Chand (*mahājan*)	

1815/1758	Mozi Ram s/o *Sah* Ratan Chand	Mozi Ram s/o *Sah* Gujar Mal
1817/1760	*Sah* Mozi Ram (*mahājan*)	Sanghi Gujar Mal (*mahājan*)
1818/1761	*Sah* Mozi Ram	
1819/1762	Faqir Das (*mahājan*) *Sah* Raja Ram (*mahājan*)	*Sah* Kirat Ram (*mahājan*)
1820/1763	*Sah* Chiman Ram (*mahājan*)	
1821/1764	—	*Sah* Lakshmi Ram Natani
1823/1766	Nonand Ram (*mahājan*)	—
1824/1767	Likhmi Ram Natani	—
1828/1771	Sanghi Mozi Ram (*mahājan*)	
1829/1772	Sanghi Mozi Ram	
1830/1773	Lachmann Ram (*mahājan*)	Sanghi Mozi Ram (*mahājan*)
1831/1774	Sanghi Mozi Ram (*mahājan*)	Sanghi Mozi Ram (*mahājan*)
	Pargana Niwai	
1721/1664	—	Bhayya Bansidhar
1765/1708	*Sah* Man Rūp	Bhayya Bhagoti Das
1780/1723	*Sah* Sambhuham (*mahājan*), Sanghi Daulat Ram	
1781/1724	*Sah* Sambhu Ram (*mahājan*) Natani Sanghi Daulat Ram	
1782/1725	*Sah* Sambhu Ram (*mahājan*) Sanghi Daulat Ram	
1784/1727	Niwaji Lal Kayath	
1785/1728	Niwaji Lal	
1789/1732	Muskas Chand Khatri	

1793/1736	*Sah* Roop Chand (*mahājan*)	
1794/1737	*Sah* Rūp Chand	
1795/1738	Chaudhurī Bakht Ram Khivsi (*mahājan*)	
1796/1739	*Sah* Malook Chand s/o Rūp Chand *Sah* Rūp Chand s/o Bimal Das (*mahājan*)	
1797/1740	*Sah* Nathu Ram (*mahājan*)	
1798/1741	*Sah* Nathu Ram (*mahājan*)	
1799/1742	Nihal Chand (*mahājan*)	
1800/1743	*Sah* Nihal Chand (*mahājan*) Tirlok Chand (*mahājan*) Bhayya Niwaji Lal	
1801/1744	*Sah* Trilok Chand (*mahājan*)	
1802/1745	Bhayya Chhajju Ram	
1803/1745	Chajju Ram Kayath	
	Pargana Malpura	
1771/1714	Ram Singh	Anūp Chand Chabra
	Ram Singh Luharya	Anūp Chand
1773/1716	Ram Singh Malūk Chand (*mahājan*) Wali Ram Natani	
1774/1717	Malūk Chand (*mahājan*)	Sanghi Baks Ram
1775/1718	*Sah* Malūk Chand Baks Ram	
1776/1719	*Chaudhurī* Fateh Ram *mahājan*	*Sah* Nand Lal

1777/1720	Bhayya Chabil Chand	*Chaudhurī* Fateh Ram
1779/1722	Sanghi Laxmi Das (*mahājan*)	Purohit Tulsi Ram
1780/1723	Lakshmi Das (*mahājan*)	Purohit Tulsi Ram
1781/1724	Lakshmi Das (*mahājan*)	Purohit Tulsi Ram
1782/1725	Sanghi Lakshmi Das (*mahājan*)	Purohit Tulsi Ram
1783/1726	*Chaudhurī Sah* Bhayya Ram (*mahājan*)	
1785/1728	*Sah* Bayya Ram (*mahājan*)	Sanghi Lakshmi Das
1786/1729	*Sah* Maya Ram (*mahājan*)	Lakshmi Das (*mahājan*)
	Sah Bhayya Ram (*mahājan*)	Jagjiwan Das (*mahājan*)
1787/1730	*Sah* Maya Ram (*mahājan*)	Sanghi Jagjiwan Das (*mahājan*)
1788/1731		Sanghi Jagjiwan Das (*mahjāan*)
	Sah Bhayya Ram (*mahājan*)	*Sah* Hans Raj
1789/1732	*Sah* Bhayya Ram (*mahājan*)	*Sah* Rūp Chand
1790/1733	*Sah* Bhayya Ram (*mahājan*)	*Sah* Rūp Chand
1791/1734	*Sah* Roop Chand (*muhājan*)	*Sah* Udai Ram (*mahājan*)
1792/1735	Sanghi Kishan Das s/o Jagjiwan Das (*mahājan*)	Purohit Akey Ram
1793/1736	Sanghi Kishan Das (*mahājan*)	Purohit Akhey Ram
1794/1737	Sanghi Kishan Das (*mahājan*)	Purohit Akhey Ram
1798/1741	*Sah* Gujar Mal (*mahājan*)	Purohit Akhey Ram
1799/1742	Sanghi Kishan Das (*mahājan*)	Purohit Akhey Ram
1800/1743	Sanghi Kishan Das (*mahājan*)	*Sah* Khushal Chand *amīn*

1801/1744	*Chaudhurī* Moti Ram (*mahājan*) *Sah* Surat Ram	Purohit Akhey Ram Brahman
1805/1748	Sanghi Lal Ram	Akhey Ram
1807/1750	Jagdish s/o Akhey Ram	Purohit Ram Kunwar s/o Akhey Ram Brahman.

CHAPTER IX

The Jagir System during the Evolution of the Jaipur State

As has been explained in Chapter I, the greater part of what later became the Jaipur state was held in the 17th century by a number of *jāgīrdārs*, most of whom were not Rajputs. *Jāgīrs* assigned to the *rājās* of Amber were not permanent at the beginning, except for the *waṭan*, which was very limited in extent. During the first half of the 18th century a deliberate attempt was made by Jai Singh Sawai to extend his *waṭan* lands by getting his ordinary *jāgīrs* assigned in the same area and contracting for other adjoining territories. His successors continued the policy until the Marathas imposed their dominance in 1750, and put the expansion of the state to an end.

Since Jai Singh was dealing with Mug͟hal territory, he first tended to adhere to those forms of administration which had been established by the Mug͟hal Government. But as his position became more and more autonomous, he began to introduce innovations with greater freedom.

For one thing, Jai Singh distributed the areas under his control among his own officers by way of sub-assignments and farms. These assignees were selected either from his own clan or from local magnates of particular localities.[1] The share of people not belonging to the ruler's clan was far smaller than those of the Kachhawahas. Very little land was assigned to

1. *Lakhtang* dated *VS* 1796/1739 of Ram Singh Shekhawat taking *ijāra istamrārī* and *zamīndārī* of *pargana* Fatehpur (2/5ths for himself and 3/5ths for Sheo Singh) for Rs. 60,000/- from the beginning of *VS*. 1796/1739. He agreed to enforce the authority of the Mahārāja in *qaṣba* Fatehpur from *VS* 1797/1740, and not to allow the '*āmal* (revenue collection) of Rathors and Qayamkhanis.

Muslims, for example ; not a single piece was assigned to them in some of the *parganas*.[2] In *paragana* Chatsu, a total number of 87¾ villages, assessed at Rs. 2,35,415/- per annum, were sub-assigned by Jai Singh Sawai.[3] The holdings in the sub-assignments were as follows :

A : *Jāgīrs* held by Hindus worth *tan* Rs. 2,34,315/- (87¼ villages)

B : *Jāgīrs* of Muslims worth *tan* Rs. 1100/- (½ village)

The breakdown of the Hindu recipients of *jāgīrs* was :

Name of clan	*Nominal Assessment (Tan) (in rupees)*	*No. of Villages*
I. Rajputs	2,21,444	80½
(a) Kachhawahas (Rājāwat etc.)	1,92,408	66
(b) Rajputs other than Kachhawahas (Gaur, Rathor, Khichhti, Nirwan)	29,036	14½
II. *Non-Rajputs*		
(a) Brahmans	6,931	3¼
(b) Kayasth	540	1
(c) Mahajan	4,450	13
(d) Bhata	950	¾

At the time of the expansion of the principality, the larger part of the territory was assigned to sub-assignees ; but as soon as the Jaipur Durbār acquired fuller control from about 1719 onwards, the area under sub-assignment began to shrink and the *khāliṣa* tended to expand.

2. *Arhsatta pargana* Chatsu *VS* 1787/1730. Also see *arhsatta pargana* Malarna and Bahatri.
3. The word signifying assessment is *tan*, for whose use see the text below.

The following Table shows the distribution of territory under *khāliṣa* and *jāgīr* under Jai Singh Sawai in different *paraganas* in different years.[4]

Percentage of villages under different categories

Pargana	*Year*	*Khāliṣa*	*Jāgīr*	*Miscellaneous*[5]
Malarna	1717	33	66	1
	1741	74	23	3
Chatsu	1708	21	70	9
	1724	66	28	6
	1741	62	33	5
	1748	46	48	6
Dausa	1719	83	12	5
	1741	33	56	1
Bahatri	1665	37	62	1
	1718	18	81	1
Amber	1715	17	70	13
	1724	43	47	10
	1734	40	45	15
	1744	32	42	26

In 1730, the proportion of revenue in *khāliṣa* to the total revenue in *parganas* Malarna and Chatsu was 90% and 63% respectively. Soon after the death of Jai Singh Sawai, there was a reversal of the process, and large territories were distributed in sub-assignment.[6]

The *jāgīrs* or sub-assignments were granted normally in lieu of pay. There were some *jāgīrs* of a miscellaneous kind, which owing to their comparative insignificance do not deserve more than a passing notice. The *ghar baitha ki jāgīr* covered assignments that were either for a specified term or, more usually, for the lifetime of the holder, lapsing on his death to the

4. This Table has been prepared with the help of the *arhsattas* of different *parganas* of the different years.
5. The miscellaneous category includes the villages held under *punya-udik, in'am, peshkash* and *ta'alluqadārs*.
6. *Lakhtangs* of different *parganas*.

state. These assignments were usually renewed in favour of the heir, on payment of a *peshkash*, and were sometimes specified as a hereditary assignment.[7] This particular kind of *jāgīr* was granted to those who had rendered outstanding service to the state either in the army or somewhere else. Usually, the holders of such grants were not bound to maintain contingents as holders of *tankhwāh jāgīr* were. At the extreme end of the spectrum were *jāgīrs* of a very lowly kind, such as those granted to persons posted to watch locusts (*khabardārī tiddi*).

The bulk of the sub-assignments or *jāgīrs* was given in lieu of salary ; and these were, therefore, known as *tankhwāh jāgīr*. Holders of such *jāgīrs* had to maintain a number of *rāsī* or *rās* (Pers. *rās* : horses), *dol* (Pers. *sawār*), and *bandūkchi* (Pers. *bandūqchī*, musketeers). The *sanad* was not issued till the horses were actually obtained and presented at the office of *baksi* (Pers. *bakhshī*). There was no *manṣab* or grading of *jāgīrdārs* presumably because the grant of *manṣab* was an Imperial prerogative. The salary of a *jāgīrdār* was generally fixed on the basis of the size of the contingent he maintained. The quotas were irregular and the emoluments of a cavalier were reckoned between Rs. 400/- and Rs. 650/- net annually (see Tables 1 and 2), depending upon the quality of the horses (of different breeds like *Turki*, *Tāzikalān*, *Tāzī*, *Yābu*, *Jangalah*)

As an illustration of the pay allowed per cavalier we may cite the case of Bishan Singh s/o Daulat Singh. He held the *tan* or salary of Rs. 10,000/- for 18 horses (*rās*) and obtained assignments in different *parganas* as follows :[8]

(i) Village Rajwas (entire), *pargana* Niwai, *tan* Rs. 8000/-.

7. *Chiṭṭhī Miti Pos Vadi* 10, *VS* 1801/1744.
8. Miscellaneous papers dated VS 1777/1720, *dīwān-i-Ḥuzūrī* R.S.A. Bikaner. Also see *Terīj sawār* from 1768/1711 to 1777/1720 *dīwān-i ḥuzūri*. Also see *Yaddāsht fahrist* Mina *gāon* and *pattadār zilhija hijri* year 1083 ; *Muwāzana pargana* Lalsot, *Yaddāsht* 1730.
Bijai Naih s/o Syo Singh, Grandson of Banwari Singh *Sultanot* was assigned a *jāgīr* worth Rs. 9000/- for *rās* 18, three remitted (*rās* 2 and musket 1); balance 15. Hence a payment of Rs. 600/- per annum per cavalier (=Rs. 50/- per month per horse) was made.

(ii) Khempura situated in Paharia, *pargana* Chatsu, one entire village, *tan* Rs. 1600/-.

(iii) Arar ka Khera, Chatsu (total *tan* 600/-), shirkat (shared village), *tan* Rs. 400/-.

The average emoluments per horse work out at Rs. 46.29 per month ($=10,000 \times \frac{1}{2} \times 1/18$).

Some of the *jāgīrdārs* were *naqdī*, that is, they were paid in cash directly from the state exchequer. From the salaries paid in cash to *naqdīs*, there were made a number of deductions such as *ponanni* ($\frac{3}{4}$ anna per rupee, which comes to roughly 5%).[9] On the other hand, since a *jāgīr* was given in lieu of cash salary, it was essential that it should yield at least as much revenue as the amount to which the holder was entitled.

An entire village (*dar-o-bast*) could be assigned to a single assignee ; but a part, such as *vat* 1 (one-fourth), *vat* 2 (one-half), *vat* 3 (three-fourths), could also be assigned.[10] In the latter case, the village could be held jointly (*shirkat*) by four or five or more assignees ;[11] and a part of it might be retained for the *khāliṣa* at the same time. The *tan* (salary assignment) of village Khizuria, for example, was shared by four persons :—

	Tan	*Share*
(I) Chain Singh	Rs. 1000/-	$\frac{1}{4}$ village
(II) Daulat Singh	Rs. 500/-	$\frac{1}{4}$,,
(III) Mukund Singh	Rs. 300/-	No share
(IV) Sangram Singh	Rs. 2500/-	$\frac{1}{2}$,,

It is not clear why the revenue of $\frac{1}{4}$th of the village should be Rs. 500/- in one case, Rs. 1000/- in another, and 1,250 in yet another.

Sometimes, the area of a small assignee is not mentioned against his salary. It is not clear whether he was paid by the higher assignee or by the state. The area and the nature of a *jāgīr* (i.e. whether the *jāgīr* to a person is assigned along with other cesses like *bhom*) was clearly specified so that a *jāgīrdār* might not realize what was not assigned to him.[12] Sometimes the

9. *Arhsatta mujmil, pargana* Chatsu *VS* 1746/1689.
10. *Arhsatta pargana* Chatsu 1730 ; *Arhsatta pargana* Malarna 1 730 ; *Muchalka Mangsir Sudi* 7, *VS* 1777/1720.
11. *Ibid.*
12. *Chiṭṭhī Miti Chait Sudi* 9, *VS* 1754/1697, *pargana* Amber.

assignment was made along with the rights of "tillage and plough" *vih-sudha* and *hal-sudha*), i.e., presumably to let the *jāgīrdār* have more control over cultivation.[13]

The *jāgīrdār* or sub-assignee earned his income from his *jāgīr* mainly from the land revenue. He appears to have possessed considerable authority for assessing and collecting the land revenue. His share ranged from one-half to one-sixth of the produce in different areas. For example, in village Sahipur *pargana* Jaitpur in 1724, the share of the *jāgīrdār* was 50% of the produce while the rest was shared by the *ra'iyatī* and the state, at 36.15% and 13.85 respectively.[14] The land revenue realized by the *jāgīrdār* was in this case nearly 64% of the produce, of which amount he paid about 2/9ths to the state, keeping the remainder for himself.

The *tan* or assessed income from a sub-assigned village was computed on a yearly basis. It seems that since *tan* (meaning literally *tankhwāh* or salary) was invariably equal to the *jama'* (assessed income) of the *jāgīr*, it is here used in the sense of *jama'* as well, and, therefore, may be taken to mean standard assessment. The exact method of computing the *tan* is not known, but it would appear that it was fixed on the basis of the cultivable land, (determined after deducting from the total area unassessable land such as *bastī* (under habitation), *rāha* (road), *ṭālāb* (tank), *sīr* (untaxed), *nala* (drain), and *pahār* (hill).[15] We

13. *Parwāna* dated *Sawan Sudi* 5, *VS* 1778/1721.
14. *Arhsatta pargana* Jaitpura, *VS* 1781/1724. Also see *Chiṭṭhī Miti Baisakh Sudi* 10, *VS* 1736/1679.
15. *Muwāzana* of different *parganas*.
In *ṭappa* Shahpura, *pargana* Jaipur, a new village Ajitgarh was founded. It comprised the area of two villages, and the *tan* was fixed after taking into consideration the area of cultivable land. The details are as follows :

(a) Village Kansubhopura measured 2000.0.0 bīghas

Uncultivated	*area*	*Cultivable area*
daugase bīgha	800.0.0	1040.0.0
tiba bīgha	100 0.0	
basti bīgha	60.0.0	
	960.0.0	

Tan amounting to Rs. 3000/- was fixed upon the cultivable area.

know that the Mughal administrators were always interested in obtaining full particulars for one or more years regarding the areas under cultivation, crop pattern, etc.; and the papers regarding *ḥāl-i ḥāṣil* of various *parganas* for the purpose of fixing the *jama*ʿ had to be sent to the headquarters. The same practice, it seems, continued in Eastern Rajasthan.

Sometimes, the *tan* of a village could be raised arbitrarily by a stroke of the pen thus leaving a sharp gap between the actual income and the assessed income.[16] The *tan* of previous years (*sābiq*) and the enhanced *tan* (*afzūd*) are often mentioned separately in our documents.[17]

The *jāgīrdār* was entitled to a *ṭalab* (claim) from the *rāj* in case the actual collection was less than the standard assessment. In such cases, the deficit was either compensated by cash payment from the state treasury or by an additional *jāgīr*. Further, the assignee was entitled for a *takhfīf* (reduction) as soon as he presented a strong plea for the same. In some cases, such as natural calamity, the claim was accepted by the state.[18]

On the other hand, if the state found justification for an increased (*gunjā'īs*) *ḥāṣil* (actual collection), either the *tan* was arbitrarily increased or a part of the *jāgīr* deemed equal in income to the increase was resumed to the *khāliṣa*.

In some of the documents we come across an 'eight-monthly *jāgīr*' (*jāgīr atha māha*). The exact nature and purpose of converting the yearly *jāgīr* (*sālina*) into eight-monthly is not

---

(b)	Village-Andesri total *bīgha*	2000.0.0	
	dugar (way) bigha	300.0.0	1500.0.0
	kharado bīgha	100.0.0	
	bastī bīgha	100.0.0	
		500.0.0	

Nett cultivable area *bīghā* 1500.00 upon which a *tan* worth Rs. 3500/- was fixed.

16. *Parwāna Asoj Vadi* 7, *VS* 1779/1722. The *tan* of village Biharipur *pargana* Chatsu was straightaway doubled from the *syalu* of *VS* 1787/1730.
17. *Arhsatta parganas* Chatsu, Malarna and Naraina 1730.
18. *Muwāẕana pargana* Lalsot.

known.[19] It is possible that the purpose of giving the eight-month *jāgīr* was to extort the same obligations from the assignees as from the person granted a twelve-month *jāgīr*, so that the members of the preferred castes or communities were placed at a relative advantageous position. It would be seen further that the monthly proportions were not arithmetically exact, such as one finds in Mug͟hal documents concerned with the *jama'ḥāṣil* ratios which were sometimes formal or arbitrary. Thus eight-monthly *jāgīr* was equal not to 2/3rd, but to 10/16th (5/8th) of the whole *ṭalab*. The figure for an eight-monthly *jāgīr* is invariably arrived at by multiplying the half or six-monthly figures by a factor of 1.25. It is, however, interesting to note that in Mug͟hal cash salary schedules for 500 *ẓāt* and less, the eight-month proportion is similarly fixed, as may be seen from the table given in the *Siyāq Nāma*.[20]

The *tank͟hwāh jāgīrs* were always subject to renewal. Every year the *sanad* had to be produced for inspection (and

19. See the 'Report on the Administration of the Jaipur State' for *VS* 2004 (Sept. 1947 to Aug. 1948), p. 34, which shows that in the deed of grant of each *jāgīr*, the *tan* or estimated annual revenue of the *jāgīr*, was entered, as also the *qarār*, i.e. the number of months during which the quota of *sawārs* and footmen required from a particular *jāgīr* was to be provided. This quota was calculated on an assumed income obtained by multiplying the *tan* by 12 and dividing it by the number of *qarār*. For each five hundred rupees of the assumed income, the *jāgīrdār* had to provide one *sawār* and for each sum of Rs. 100/- or fraction thereof over and above Rs. 500 or any multiple thereof, he had to provide one foot-man, five of the latter being usually regarded as equivalent to one *sawār*. The quota to be provided was entered in each deed of grant, which also contained entries of the number of horse men exempted from service either by customs or by special favour of the ruler. The obligations to provide cavalrymen (*sawārān*), infantry, etc., was commuted for cash in 1925. The highest class of *jāgīr* was *manṣab*, the holders of which had to pay commutation dues (*naqdī*) at the rate of Rs. 22/ Rs. 8/- per horse per month. The other *qarārs* are ten-monthly, eight-monthly, seven-monthly and six-monthly, the last of which was the most common. The commutation rates for these are Rs. 18/ Rs. 12/-, Rs. 16/Rs. 8/- Rs. 13/8- and Rs. 11/4/- per horse per month respectively.

20. *Siyāq Nāma*, printed by Naval Kishor, 1879, pp. 272-77. The *Siyāq Nāma* was compiled in the 40th R.Y. of Aurangzeb. I owe this information to Prof. Irfan Habib.

confirmation) except in the case of those exempted (*ma'āfī*). Failure to produce an appropriate *sanad* could lead to confiscation *kurak* (Pers. *qurq*) and resumption of the *jāgīr* to *khāliṣa*. In the case of the incumbent dying, the *jāgīr* was generally assigned to either his son or his close relations.[21] In the case of a minor heir, a temporary substitute was found and the *jāgīr* given to him, if he was prepared to serve the *rāj* upon the same conditions.[22]

The assignments were often subject to temporary resumption for the satisfaction of *muṭāliba* (state's financial claim upon him),[23] or failure on the part of *jāgīrdār* to discharge the obligations imposed upon him, e.g., absence from duty,[24] or failure to produce the required number of horses at the time of inspection or to get them branded,[25] or not producing the required *sanad*, or *musaadatī*[26] (unpaid loans), or *tafāwat khurāk ghora*[27] (fodder for horse), or payment of the *peshkash* (offering).[28] Sometimes a part of the *jāgīr* was taken into *khāliṣa* to realize the amount of the *muṭāliba*.

Unlike the other Mughal *jāgīrdārs*, the Jaipur Rājas began to assert the right of sovereignty and to emphasize that the tenure of their sub-assignees (*jāgīrdārs*) depended on service and loyalty to them personally. But a study of the contemporary evidence suggests that, in their turn, the *jāgīrdārs* began in course of time to claim their tenure by right of inheritance and could not be easily dislodged. The Jaipur rulers held them duty-bound to pay *peshkash* while entering the service of the

21. *Muwāzāna pargana* Lalsot.
22. *Muwāzana pargana* Dausa.
Also see *Yaddāsht Miti Mah Vadi VS* 1772/1715. Subha Ram, two years old, was granted the *jāgīr* of his deceased father worth Rs. 500/ annually. Kishan Sihgh was accepted as the substitute of the minor heir.
23. This would parallel the Mughal practice. See Irfan Habib, *Agrarian System of Mughal India*, p. 269.
24. *Awārija mutāliba pargana* Bahatri *VS* 1749/1692. *Chiṭṭhī Miti Sawan Vadi* 5, *VS* 1749/1692.
25. *Awārija muṭālibo*, op. cit.
26. *Ibid.*
27. *Ibid.*
28. *Ibid.*

Mahārāja and, in times of peace, pay proper homage to the Jaipur Durbār. The Jaipur rajas also reduced many of the bigger chieftains (territorial *thikānādārs*) such as those of Manoharpur and Khandela, to the position of *jāgīrdārs*. The Shaikhawatis and other *thikānādārs* also accepted service in the Jaipur *rāj*.[29] (See Table I). Wills holds that :

> they were in fact, in the proper sense of the term, 'feudatories' of the Mahārāja of Jaipur. They held their land under a grant from the state to which, by established usage, liability to render service as well as to assessment was attached. They were not *jāgīrdārs* in the sense that a specified area was assigned to them for the maintenance of troops. They were *jāgīrdārs* in the sense that a specified deduction from the assessments of their landed property was assigned to them for service to the Mahārāja.[30]

But the deliberate policy of the Jaipur Durbār seems to have been to convert these *thikānādārs* into regular *jāgīrdārs* as may be seen in the case of the Shekhawatis. By so doing the Durbār tended to extend its absolute sovereignty over them. The *thikānādārs* of Shekhawati originally claimed a special status since they considered themselves autonomous chiefs. On the other hand, the Jaipur state, at the Wills Enquiry, claimed that the *thikānādārs* of Shekhawati were not autonomous chiefs, and their status was not different from the rest of the *jāgīrdārs*.[31] (For the position of Shekhawatis and their relations with Jaipur Durbār see Chapter X.)

29. *Yaddāsht Mangsir Vadi VS* 1773/1716.
Terij Shekhawat Ki VS 1766/1711. In this *terij* the names of the Shekhawat *jāgīrdārs*, the number of horses to be maintained by an individual, and the salary per annum (sometimes *naqdī* and sometime through assignments) are given.
Also see *Fahrist Sawaran Mulajim Sarkar* (*Shekhāwat*).
30. C.U. Wills, *Land Tenures and Special Powers*.
31. *Ibid*. Also see Jackson John, '*A Reply to the Report on the Land Tenures*' *and Special Power on Certain Thikanadars of the Jaipur State*.

Table I

Terij Shekhawat, Pargana Amber

1709

Asāmī (*Name of the jāgīrdār*)	*Dol* (*number*)	*Rāsī contingent*	*Yearly tankhwāh* (*in rupees*)
Gaj Singh etc.	5	49	15,700
Chand Singh	5	11	6,000
Jodh Singh	2	10	5,500
Jaskaran	3	9	5,100
Jhujhar Singh	1	11	7,500
Daloo Singh	1	4	2,000
Dev Singh	1	4	2,000
Daulat Singh	7	81	45,000
Bhao Singh	1	22	12,000
Bhim Singh	1	3	1,500
Ran Singh	1	5	2,500
Ram Singh	1	3	1,600
Sukh Singh	1	3	1,600
Sawat Singh	1	5	3,100
Akhey Singh	1	6	3,500
Amar Singh	1	14	8,000
Inder Singh	1	5	2,500
Jaswant Singh	2	6	3,000

Average payment per horse Rs. 510/-per annum (=Rs. 42.50 per month)

Table II

Terīj soldiers of *jāgīrdārs* as 1807/1750

Asāmī	*Dol* (*number of persons*)	*Rās* (*number of horses held*)	*Yearly Rs.* (*Salary per annum*)
Anoop Singh	1	1	600/-
Arjun Singh	1	3	1500/-
Ajab Singh	1	4	2300/-
Udol Singh	1	2	1100/-
Amar Singh	1	1	600/-
Ajab Singh Bhati	1	1	600;-
Amar Singh T	1	1	600/-
Ajab Singh	1	1	600/-
Akhey Singh Rathor	3	3	1600/-
Arjun Singh	1	2	1100/-
Abhay Ram	3	3	1600/-
Azmat Khan Palhani	2	2	1200/-
Anand Ram	2	2	1200/-
Anand Ram	2	2	1100/-
Abhay Ram T	1	1	500/-
Kushal Singh	4	4	2400/-
Kishan Singh T	1	1	600/-
Kisfar Singh Rathor	1	1	600/-
Kesri Singh H	1	1	600/-
Kirpa Ram	1	1	600/-
Karam Khan Pathan	1	1	800/-
Keshri Singh	1	1	600/-
Total	32	39	22400/-

Average payment per horse : Rs. 574/- per annum (=Rs. 47.83/- per month)

CHAPTER X

The Farming System

Ijāra constituted a kind of contract and implied the farming of revenues of a village, a part thereof, a *mahāl* or *pargana* or a still larger tract. Generally, the *ijāradār* or the farmer was required to pay a fixed amount stipulated in the *paṭṭa* or document of lease. He had to give a *qubūliyat* (acceptance), affirming his obligation to abide by it. It was expected that he would protect the interests of the state as also of the *ra'iyatī*. The settlement was sometime made for an amount that was a little less than the estimated income from the area, thus leaving a margin of profit for him. We may suppose that an efficient *ijāradār* could make an overall estimate of the expected revenues from the land under cultivation before a detailed assessment have been made by the *jāgīrdār*. His vigilance and personal efforts could increase his collection beyond the estimated *jama'*.

Ijāra could be of serveral types. Broadly, however, we may classify it in two categories : the *ijāra* taken by the rulers of Amber/Jaipur ; and *ijāra* taken by private persons from the rulers in lands held by the latter in *jāgīr* or in *ijāra* of the first kind. The former, again was of two sub-types :

(i) *Ijāra* of a *pargana* or *parganas* outside the tracts under Amber/Jaipur, often taken when the Imperial (Mughal) *jāgīrdārs* (*manṣabārān-i-Bādshāhī*) or the royal servants were not able to realise their revenues.

(ii) *Ijāra* of isolated villages of the Imperial *Khāliṣa* or farming of the *jāgīr* of the Mughal *manṣabdārs*, falling within areas already controlled by the Jaipur rulers or surrounded by such territory.

We have seen in Chapter I how Jaipur acquired its status as a major state of Rajasthan as a result of the position enjoyed by its rulers in the service of the Mughal Emperor. The expansion of the territories of its rulers did not take place essentially as a result of military conquests, but due to the transfer, over a period of years, of many adjoining *parganas* by the Mughal Emperors to the rulers of Amber. These *parganas* were given sometime in *jāgīr* and sometimes in *ijāra*.

The leasing of *parganas* in particular proved to be a crucial factor in the expansion of the Jaipur state. During the process of disintegration of the Mughal Empire, the Kachhwaha rulers strengthened their position by their willingness to take large areas on farm. The crisis in the *jāgīr* system of the Mughal Empire seems to have led to much greater recourse on part of the Mughal nobles or *manṣabdārs* to the practice of giving their *jāgīrs* on farm to the rulers of Jaipur, thereby assuring themselves of a stable income. The practice of revenue-farming began to increase after the death of Aurangzeb and Bahadur Shah and in the reign of Farrukh Siyar became noticeably widespread. Efforts to take many *parganas* on lease were especially made by the Amber Durbār in 1712 (see Chapter I).

After 1712, the *ijāradārs* had still greater opportunities to obtain extension of the term for which *ijāras* (contracts) were originally given. The tradition of short-term farms and frequent changes began gradually to disappear. Formerly, the *ijāras* were temporary ; but after some time the Maharaja of Amber acquired some of the *ijāra* land permanently or for life. Whether holding in *jāgīr* or *ijāra*, Jai Singh Sawai acquired full authority and complete political jurisdiction throughout the territory under his control. He managed many of these *parganas* through leases or subordinate *ijāradārs*.[1] Such subordinate *ijāradārs*, on their part, were on their way to becoming landed chiefs, during the period of our study, just as their ruler tended to become a permanent holder of his *ijāras*.[2] It appears that the new *ijāra* system had this great attraction for the holder, as compared with the assignment, that it gave him some hope

1. *Arhsattas, pargana* Lalsot and Malarna, *VS* 1787/1730.
2. C.U. Wills Report on '*The Land-Tenures and Special Powers*,

of prolonged possession. The list of the *parganas* taken on *ijāra* by Jai Singh Sawai shows that he occupied by this means an area much exceeding what he received through *tankhwāh jāgīrs*, or ordinary assignments, from the Mughal Emperor.[3]

Efforts were also made to secure *ijāras* of the *maḥals* assigned to Mughal *manṣabdārs* or nobles close to the Mahārāja's *waṭan*. Due to the weakening of administrative control during the first half of the 18th century, particularly after 1719, the Mughal *manṣabdārs* could not realise their revenues from *maḥals* situated around Amber. They began to transfer their assignments to the Rajput rulers mainly in the form of *ijāra* or contractual arrangements.

On the other hand, the *mutasaddīs* (officials) and other influential persons at the Imperial Court were offered gifts and inducements by the agents of the Mahārāja so that they might prove helpful to him in securing *ijāras* from Imperial nobles.

There was a good deal of bargaining between the *jāgīrdār* and the revenue-farmer (*ijāradār*) regarding the terms and conditions of the *paṭṭa*. But increasingly it was the farmer who tended to have the best of the bargain. For instance, the *jāgīr* of Saiyid Khan, in *pargana* Toda Bhim (*dar-o-bast*) was sought to be taken in farm by Jai Singh Sawai. The *paṭṭa* was prepared and *qubūliyat* was sent, but later on the Maharaja returned the *paṭṭa*, and did not send the *māl-ẓāminī* (surety) to the *jāgīrdār*, arguing that there was scarcity of rain and some of the villages of the *pargana* had already been given on *ijāra* to others by the *jāgīrdār*. Consequently, after long correspondence, the Maharaja prevailed and the *pargana* was taken on *ijāra* by him on his own terms.[4]

The Mahārāja always tried to secure the *ijāra* of a *pargana* in entirety (*dar-o-bast*). If a *pargana* was given in *jāgīr* in part, close to the Maharaja's *waṭan*, this usually impelled the Maharaja to seek to take over the remainder of the *pargana* also on farm.[5]

3. *Parwāna* (in Persian) of Maharaja to his *vakīl* at the Mughal Court, dated 9th *Ramzān*, 51 Regnal Year (1707) ; also see *Vakils* Report dated 1712.
4. '*Arẓdāsht* (Hindi), *VS* 1768/1711 from *dīwān* Bhikari Das to Maharaja of Amber.
5. *Arhsatta pargana* Lalsot *VS* 1747/1690.

The second type of *ijāra* comprised farms given by the Maharaja of Jaipur to his *jāgīrdārs* or subordinate officials. The farmers were usually either persons of local standing or deemed capable to hold their own amidst turbulent conditions. The unit of the farm (*ijāra*) could be very small, indeed. A village or a part of it or a particular tax such as *ghāscharāī*, *sadrī* (iron), *dastūr chaudhurī qānūngo*, or all taxes (*kul ḥabūbāt*) could be farmed out by the Mahārāja for a year or less than a year, or for a single harvest.[6]

A study of the *khāliṣa* villages held by the Jaipur Mahārāja may help us in understanding the pattern of the farming practised at the time. Due to revenue farming, the size of the *parganas* did not remain constant. The fluctuation in *khāliṣa* land was also considerable. This can be seen from detailed information in documents like *arhsatta* and *muwāzana*, where villages given out under *ijāra* are distinguished from those retained in *khāliṣa* for which alone detailed accounts of tax collection are given.

The territory held in *khāliṣa*, as well as that given on *ijāra* within it, could increase and decrease from year to year and from season to season. The change in the position of villages was so rapid that a village could be *dar-o-bast* (held entirely in *khāliṣa*) one year and *shirkatī* (shared) in another year. The village or a part of it could be held in *ijāra* for one year and in *khaliṣa* for another year. A detailed study of *muwāzanā* documents suggests that the changes were so frequent that a sub-*ijāradār* could not be sure that the portion which he held in farm would remain with him for a long time.[7]

Sometimes the Maharaja used to give such portion of the *khāliṣa* in *ijāra* as lay amidst villages (or even portions of a village) assigned by him to his *jāgīrdārs*. In such cases the larger *jāgīrdār* was preferred for the position of *ijārādār* of the *khāliṣa* share.[8]

6. *Arhsattas parganas* Chatsu and Malarna, *VS* 1787/1730.
7. *Muwāzanā, pargana* Dausa and Lalsot.
8. A few instances may be quoted :

   (i) Villages Mukaliyameda *pargana* Chatsu was in the *jāgīr* of Bhoj Raj Hamirde in 1663 for a sum of Rs. 100/- only, which later on was made *khāliṣa*. Half of the village (*tan* Rs. 50/- only) was given in *ijāra* to Ghanshyam Vijai Ram for a sum of Rs. 37.50 only.

Villages held unshared (*dar-o-bast*) in the *khālisa* were also given in *ijāra*, but the practice was not common during the period under study.[9]

A village or a part of it was also given in *ijāra* to its *patel* and *mahājan* (who could also hold a *jāgīr* under the *mahārāja*).[10] *Ijara* in this form was undoubtedly an old practice. The *paṭels* and *mahājans* were local officials and, presumably, also local *zamīndārs* and peasants.

From our evidence on *ijāra* some preliminary conclusions may be drawn. First, the Durbār preferred to give the *khālisa* share on *ijāra* to one of the big or influential assignees already holding part of that *pargana* ; but if all the co-sharers were small, the contract could be given to an outsider.[11] Secondly,

---

(ii) Village Khejri *Buzurg* (*pargana* Chatsu

tan (assessed income)	Rs. 5,100/-
jāgīrdārs' share	Rs. 5,025:-
(a) Puran Mal *mahājan*	Rs. 3,750/-
(b) Dev Singh Sikarwal	Rs. 1,275/-
khālisa	Rs. 75/-

The *māl* (revenne) of the limited *khālisa* share was given in *ijāra* for the whole year to Puranmal *mahājan* (*jāgīrdār*) according to the *qubūliyat* 1729, for a sum of Rs. 50/- only.

khārīf Rs. 25/- *Rabī*' Rs. 25/-

(iii) Village Chimanpura *pargana* Chatsu (¾ village was in *khālisa* and ¼ in *jāgīr*)

tan	Rs. 4,500/-
In *jāgīr* of Shyam Singh Vijai Singh (*kumbhani*)	Rs. 1,850/-
khālisa	Rs. 2,650/-

Māl of the *khalisa* was given in *ijāra* to Shyam Singh Vijay Singh *kumbhani* according to the *fard* of *dīwān* Rao Narain Das Kirpa Ram dated 1730. According to the *lakhtang*, 1730, *māl-zāminī* (security) of the *patel* of the same village was asked for. The *ijāra* was for six months for the sum of Rs. 1,325/- only.

(iv) Village Kishanpura *pargana* Chatsu *tan*	Rs. 800/-
Jāgīrdārs share	Rs. 200/-
(a) Hirdaya Ram	Rs. 100/-
(b) Khushal Singh	Rs. 600/-
khālīsa	Rs. 600/-

The *māl* of the *khālīsa* was given in *ijāra* to Gheena *mahājan*, resident of Shahpur *pargana* Amber in 1729 for the sum of Rs. 300/-.

9. *Arhsatta pargana* Chatsu, *VS* 1787/1730.
10. *Ibid.*
11. *Ibid.*

the amount for which the *ijāra* was given was usually lower than the *tan* or officially assessed income of the area (though there are other instances when the amount of *ijāra* was equal to the assessed income).

The period of the *ijāra* given by the Durbar was not uniform. It ranged generally from six months to one year, as stipulated in the terms of the *paṭṭa*. Sometimes it could be given for two to three years but rarely for four and five years or more.[12] In such cases if half of the village was in *kh̲āliṣa* and half in *jāgīr*, the *ijāra* was usually given for a term of six months. Certain standard instructions were issued for setting terms of the *ijāra*.[13]

The *ijāradār* was sometimes granted a reduction (*takhfīf*) in the amount due in instalments contracted under the terms of the *paṭṭa*. Only a convincing argument (such as scarcity of rain or collection of a smaller *haṣil* than assessed) could help him to win concessions in revenue ; and his statements sometimes had to be verified by the *chaudhurī*, *qānūngo*, *paṭwārī* and *qāẓī*.[14]

The *ijāradārs* were invariably required to produce a surety who was generally a *patel*, a *mahājan* or a *sahūkār*.[15] The

12. Bond, *VS* 1785/1728. *Purohit* Arat Ram to Mahārāja informing him about the surety of an *ijārdāar* for a period of 5 years *VS* 1784-1788. Also see *muwāzana kh̲urd* regarding the *ijāra* of *pargana* Khohri for a period of ten years. But such instances are not common during our period. According to the *arẓdāsht* of Pancholi Jagjiwandas (Mahārāja's *vakil* at Mugh̲al Court) dated *Miti* 6, *Asarh Vadi* 17, Tuesday *VS* 1771/1714, the *maḥal* of Khohri was declared to be in the *watan* of the Mahārāja. This fact was emphasised at the time when Churaman Jat was trying through Khan-i-Dauran to take the *zamīndārī* of *pargana* Khohri after paying a large *peshkash*. The *vakil* claimed that the *pargana* should also not be assigned to anyone else either. (My attention to this document was drawn by Miss Sumbul Halim).
13. *Lakhtang* of Daya Ram *mahājan* about *pargana* Malarna, *VS* 1769/1712.
14. *Arhsatta*, *pargana* Lalsot, Naraina, Malarna, etc ; *Arzdāsht dīwan* Bhikari Das to Mahārāja *VS*17 68/1711.
    Also see *muwāzanas* of various *parganas*,
15. The *sāhūkār* or *bohra* (money-lender) used to stand as a guarantor on behalf of an *ijāradār*. The *sahūkārs* had quite often agents or partners in important *parganas* and towns of Rajasthan and other

mahārāja could refuse to grant a *paṭṭa* of *ijāra* without the production of a proper surety. Sometimes he pressed the *ijāradār* to submit *māl-i-ẓāminī* (security) through a particular *sahūkār*. The protection of the *sahūkār* was the direct responsibility of the *jāgīrdār*. There are instances when *māl-i-ẓāminī* or the security (in cash) of a *sahūkār* was looted in the jurisdiction of a *jāgīrdār*. The *jāgīrdār* concerned was required to return the money. Sometimes the amount was restored to the *sahūkār* by the State which later on recovered it from the *jāgīrdār*.[16]

I have set out in the Appendix of my paper published elsewhere a detailed table indicating the proportion which the assessed income from the lands in the *k͟hāliṣa* given in *ijāra* bore to the income from the entire *k͟hāliṣa* land in *parganas* Amber, Chatsu, Bahatri (Baswa), Dausa, Malarna and Lalsot for the period 1665-1749.[17] Although as of now there are long gaps in the information for each *pargana*, further scrutiny of the *arhsattas* is likely to fill many such gaps. Incomplete as the information is, it strongly suggests that around the years 1733 and 1734 the proportion of *ijāra* lands began to increase, *pargana* Malarna being the only exception. The increase in other *parganas* is so substantial that one is constrained to pose the question whether the Jaipur Durbar had begun to face serious financial difficulties from the early 1730s onwards. This alone can account for the fact that the Durbar was giving away larger and larger chunks of the *k͟hāliṣa* lands to farmers. Such a policy, once adopted, was self-defeating because the farming of revenue both immediately and ultimately must adversely affect the prosperity of the cultivator. But the decline in the Durbar's own capacity to realise revenue at a constant or higher scale might have persuaded the Durbar to resort more and more to farming in order to get as much as possible. Thus there would have begun to operate a disastrous cycle of cause and effect.

places such as Delhi, Agra and Ajmer. They issued *hundīs* (bills of exchange).

16. *Chiṭṭhīs* of *pargana* officials to *diwān*.
17. S.P. Gupta : 'Ijāra System in Eastern Rajasthan (c. 1650-1750)', *Medieval India-A Miscellany*, Vol. II, pp. 263-275,

An interesting point to note is that there are very few instances of *ijāra istamrārī* (permanent or long-term revenue-farm) during our period. But during and after the second half of the 18th century the practice of this kind of *ijāra* became common, and whole *parganas* began to be given away by the Jaipur rulers on *ijāra*. It appears that after the death of Mahārāja Ishwari Singh, the Jaipur Durbar faced grave political instability in the areas formerly held by Jai Singh Sawai. The threat from the Marathas might also have forced the Durbar to farm out those areas at lump sum fixed permanently.

The grant of short term and then long term *ijāra* by the Durbar was perhaps one form taken by the process of the rise of the territorial *thikānādārs*, who later on began to act almost as independent chiefs. Before the coming of the British they had already acquired a kind of proprietary tenure, and their territorial rights were now never questioned. Wills divides the history of the territorial *thikānādārs* of the Jaipur State into three periods. The first is the period in which their territorial rights developed. In the second phase, they rose gradually to the position of practically independent chiefs, owing an uncertain allegiance to the harassed Mahārāja of Jaipur. During the Maratha period the establishment of their territorial rights was still a point at issue. In the end, they won their way to a solid proprietary tenure, subject to the payment of a small quit-rent to the State. In the third phase, which followed the Anglo-Jaipur Treaty of 1818, their territorial rights were deemed firmly established ; but the political independence which they had acquired in the previous period was of course lost, though the British government conceded them autonomy in the management of their internal affairs.[18]

Wills in his report held the view that these *thikānādārs* were directly established by Mahārāja Jai Singh Sawai as his *ijāradārs* ; and they were "neither Feudatories who were settled upon their estate as a reward for their share in an original conquest, or who carved out that share for themselves",

18. C.U. Wills, '*A Report*' . . . op. cit.

nor again "tribal" tributories linked politically to their overlord by the ties of kinship.[19]

So to Wills, unlike Western Rajasthan, where the rise of *thikānādārs* was in reward for their share in an original conquest, in "Jaipur the basis of the tenure of their land was originally an *ijāra*." True, the *thikānādārs* of Shekhawati nevertheless contended that "they have, from their first entry upon Panachpana-Singhana (two *parganas*), always been independent chiefs".[20] Prior to their occupation of this territory the "Qaimkhanis had been there for nearly 200 years and held their territory as chieftains".[21] As the "previous proprietors or chiefs of the Panachana-Singhana area, the Qaimkhani Nawabs, were ousted by Sardul Singh who, by such ouster, stepped into their shoes and acquired the position of landed proprietor or chieftain and so enjoyed a status which had been formerly enjoyed by the Qaimkhani Nawabs. Later on all the Panachana-Singhana Sardars entered into a treaty with Jaipur, in which the Shekhawati *thikānādārs* agreed to pay tribute. They further maintained that by virtue of this treaty, they enjoyed a status different from that of many other *thikānādārs* of the state. According to Tod "The Sekhawat confederation was superior to, and independent of, the parent state."[22] Wills was of the view that "of Sawai Jai Singh's extension of his territories by means of the farming system neither Tod nor any other subsequent writer has had the least conception. All, therefore, have been equally at fault in their attempts to analyse the character of the tie which links his subordinate estate-holders with the Ruler of Jaipur."[23]

Jackson criticised Wills' theory of the origin of the *thikānādārs* by saying that the taking of *ijāra* of *jāgīrs* does not in any way affect the *thikānādār's* independent or proprietary

19. *Ibid.*
20. *The Jaipur Gazette, Extraordinary*, Vol. LVI, 19th May 1939, No. 4864. Quoted from *Vigat* no. 9, p. XIII.
21. Wills Inquiry Committee,, *The Report on Panchapana-Sighana.*
22. James Tod, '*Annals and Antiquities*', pp. 313-324.
23. C.U. Wills, '*The Report on the Land Tenures* . . . para 63. cf *Vigat* **No. 9.**

rights within that *jāgīr*.[24] The *ijārādār* was entitled to collect the revenue of the tract only ; if it was a chief's territory, the farmer would collect from the chief.[25] It did not give him any possession of the tract. Jackson argued that "if one considers the position of an *ijāradār* one realizes that it never gave him a right of sovereignty."[26]

Fateh Singh Champavat who was the Prime Minister of Jaipur and was entrusted with the task of writing a history of Jaipur by Mahārāja Ram Singh, has said : "Mahārāja Sawai Jai Singh Ji brought under him the several Shekhawat chiefs of Shekhawati."[27] While giving the boundaries of the Amber states at the time when Mahārāja Sawai Jai Singh came to the throne, he does not include Shekhawati and says, "All the Shekhawat chiefs had become independent."[28] This rather supports Jackson's view with regard to the Shekhawati, whose *thikānādārs* had as we have seen, strenuously contested Wills presumption about the origin of their rights.

Unluckily, the documents on which both sides rested their claims are not available ; one can only hope that they have been preserved. Until these turn up, however, one can provisionally say that the *thikānādārs* of the Jaipur State arose mainly out of the *ijāra* system, as Wills believed. In the Shekhawati, the position was that the chiefs already established were made *ijārādārs*, so that they at least could claim an earlier history for their *thikānās*.

24. John, Jackson *A, Reply to the Report on Wills Land-Tenures . . . .*, para 20.
25. *Ibid* , para 21.
26. *Ibid.*, para 19.
27. Champavat Fateh Singh, Preface to '*A brief History of Jaipur*', 1899, (Moon Press, Agra, 1899). Quoted from *Vigat* No. 9, prepared by Director Department of History, University of Rajasthan.
28. In this *Vigat*, Dr. Joshi under the heading, 'From the Director Desk—A Word about Papers and Events' has tried to give the details of events on Wills Report. Some extracts have been taken by me from it.

Appendix A

SELECTED DOCUMENTS

APPENDIX A-1

Raqbo Pargana Mauzabad VS. 1698/1641

This is an extremely interesting document. It provides us, among other things, with the *pargana* and village-level details of measured area (cultivable and uncultivable land), land alienated in the form of grants (*punya-udīk* or lands given for charity purposes and thus free from assessment), castes of revenue payers in the villages, etc.

Pargana Mauzabad was assigned to the Kachhwaha rulers of Amber in *tankhwāh jāgīr* from time to time. The *raqbo pargano* (literally the area of *pargana* Mauzabad) of 1641 was undertaken by the officials of the Maharaja. Those were the *shiqdār* Rao-Bharat Singh *Kārkun Sāh* Bandi Das and Than Singh and the *amīn* and *purohit* Kisori Das and Ajo Raj, who recorded the *raqba* (area) of the *pargana* and its villages after making a thorough inquiry from the *patels* and *paṭwārīs* of the individual villages.

According to this survey, *pargana* Mauzabad in 1641 consisted of 78 villages, distributed among *ṭappas* (administrative units) viz : (i) *ṭappa ḥaveli* (*pargana* headquarters) containing 40 villages ; (ii) *ṭappa* Phagui (23 villages) and (iii) *ṭappa* Jhak (15 villages). Out of the 78 villages, 9 were granted in *punya-udīk*. The account of the remaining 69 villages is contained in the *raqbo pargano* which proceeds to give information on the following items :

(i) Total measured land (Persian *ārāẓī*), (ii) geographical situation of each village with all its directions given in terms of *kos*, (iii) the names of *ra'īyat* (revenue payers), confined to the *patels* (village headman) and *paṭwārīs* (village accountants). The caste of the persons designated *ra'īyat* is also recorded.

The extensive statistics of the total measured area provide us with a possible means of estimating the extent of cultivation in 1641. The actual percentage of the cultivable area was approximately 92% in the whole *pargana*, but varied from village to village ranging between 84 and 96%. The cultivable area is relatively smaller in those villages which are said to have been founded recently. Unfortunately, the area actually cultivated is not given. This bears out the complaint voiced in Aurangzeb's *farmān* to Rasikdas (1666-7) that it had become a general practice with the revenue officials not to report the area actually sown.[1]

A majority of the *paṭels* and *paṭwārīs* in this *pargana* belonged to the castes of Jats, followed by Gujars, Brahmans, Ahīrs, *mālīs*, and Brahmans. It is interesting to note that a larger number of Brahmans were also *paṭels*, thus becoming involved in cultivation and the collection of revenue.

The document gives the number of bullocks owned by each peasant and so supplies us important data about the heads of cattle (bullocks) per peasant and the stratification among the peasants. It also informs us of peasants who had migrated from, neighbouring *parganas* or had been in service (*naukar*); among these some were previously wage-labourers.

Cultivable Area and the Ra'iyatī Castes
Pargana Mauzabad, 1641 AD.

Villages	*% of cultivable*	*% of cultivable*	*Ra'īti Caste*	*Remarks*
1	2	3	4	5
Pargana Mauzabad	91.92	8.08		
Qasba Mauzabad	91.30	8.70	Gujar and Jat 4 *paṭel*, Gujar, 1 Jat, 2 *kumbhārs*	1 Brahman, 1 *paṭel*
Dhamana	85.71	14.29	Jat	*Ra'ītī* here signifies only *paṭel* and *patwārī*
Paloo	91.17	8.83	Jat and Gujar	
Kisanpur	—	—	Jat	
Gidhani	95.96	4.04	Gujar, Jat	7-8 houses belonged to Brahman and *banias*, who cultivated the land.

1. Aurangzeb's *farmān* to Rasikdās, ed. J-Sarkar, JASB, NS, 11 (1906), pp. 223-55.

1	2	3	4	5
Jaisinghpura	66.67	33.33	Jat (*paṭel*), *paṭwārī Saraogi, bania*	Newly founded.
Sahpur	—	—	Bania *patwārī*), Jat (*paṭel*), Brahmans (*paṭel*)	Newely founded, two houses of Brahman
Chararda	—	—	Jat	——
Dogra	18.18 45.44 (pastured land)	36.36	*kharel, mali* Gujar	*Pālti* of the village, is cultivating the field.
Urseva	90	10	5 Jat (*paṭel*) 1 *paṭwārī Bania*	5 *paṭel* 1 *paṭwārī*
Thehura	86.67	13.33	Cultivated By Jat (*paṭel*) of *qaṣba* Mauzabad	——
Kakrali	*ujar*	*ujar*	——	granted in *punya-udik*.
Fagui	94.28	5.72	Brahman, *bania, gunwar* and Jat	The rest cultiv ated by *jāgīrdārs'* persons.
Choohu	92.94	7.06	Jat	——
Chekwara	84.25	5.75	Jat & Brahmans	Approx. 15 houses of Banias
Pachala	94.28	5.72	Jat & Brahmans	——
Goindee	85	15	Ahir, Jat, Brahman, *turuk* (Muslim)	——
Rampur Govindee	85	15	Jat	——
Rampur Sadi	87.50	12.50	Jat, *mali* and Gujar	——
Datuli	91.04	8.04	Jat	4-5 houses of *turuk* (*Muslims*
Kagyo	95	5	Brahman, Ahir	5 houses of Brahman, 3 houses of Ahir
Jhak	90	10	Jat	The rest by persons of *Jāgīrdār*

रकबो परगनौ मौजाबाद साल सं० १६९८

रकबो परगनौ मौजाबाद सरकार अजमेरी सुवे अजमेरी अमल माहराजाधिराज माहाराजा श्री जी सीकदार रा. भरतसिंध जी कारकून सा. बंदी दास वा साह थानसिंध अमीन प्रौहित कीसोरदास वा सा० अजो राज संवत् १६९८ साल सन १०४९ को सालि पटेल-यटवारी गांवा थे बुलाय अर पूछी लिखी तपाई का मौजा—७८

तपाहवेली		तपा फागई	तपा झाक
मौजा		मौजा	मौजा
४०		२३	१५
	असली	इजाको नवो	
	मौजा	वसो गाव मौजा	
	२२	१	

वादि पुन्य उदीक मौजा ९

तपा हवेली,	तपा झाक
मौजा ५	मोजा ४

वाकी मौजा ६९ मुकररा जीमी उनवान वीघा ६२६२८६

नीस तुक नीसा गाव गाव को रकवो डोरी मापी सुवा होई सो सही और इनाम फुटकरी डोहल्या वगेरेहो जो जीमी पावौ छो तीह सुधी

असलि जीमी वीघा ५७५६८६

खारड़ा उसर तलाव नाला खाला
गाव गैला वगेरहो वीघा ५०६००

तपाहवेली मौजा ४० वादो पुन्य उदीक

मौजा———

५

मो० ड़ोगरी कसवा
थे कोस पौण दीखण
दीसाय चाहुण चारण
नो उदीक १

मौ० चीड़ासरे कसवा
थे कोस पांच पछीम
दीसा ———
१

गंगावति खुरद
कसवा थेकोस पौण
उत्तर दीसा वेभोज—
राज चारण ने १

मौ० छापरवाड़ौ कसवा
थे कोस तीन पीछम
दीसा वे राव हरचंद
वगौ० दसौधी—१—

खेड ।कसवा थे कोस ——————
द० थारी नोरीत कणही ——————
सो कान्हा चारण वगौ० १ ०
वाकी मोजा ३५ मु० जीमी उनवान
वीघा ————
२७६१६२

असलि जीमी	खारवगो०
वगौरहे	वीघा
वीघा	———
२५१४६२	२४७००

कसवै मोजावाद वा कसवा की जीमी
गाव स्या तह सुधी वीघा ———
७००००

असलि जीमी	खारवा उसर नाला
वीघा	तलाव गाव गौला
	वगौ० वीघा ——
६४७००	५३००

कस्वे मोजावाद
वीधा ———
४६०००

असली जीमी
वीधा
४२०००
वीड़ द०
—— ———
१५०० ४०५००

खारड़ा वगो
४०००

मौ० रामपुर का अमल
महाराजी श्री मानसिंध
जी के बसे जीमी वीघा
६५००

कसवाथे कोस
सवा मुखदीसा
असलिवा—
हवालायक ६३००
खा रड़ा वगो वीघा २००
रौती जाट लोग वसै

रौती गुजर जाट

भुरा पटेल	खरथे पटेल	चोखे वेगा	देवे नायाको
सुरजन का	पीथा को गुजर	को पटेल	पटेल जाट
गुजर जाति	जाति कुतण	जाति जाजोहो	जाति ककरालो
खटाणों १	१	१	१

नेते पटेल	रामे पटेल	तेजे वीरमे को	सा० नेतसी
वांल्हा कै जाट	खेमा को कु—	पटेल जाट	पटवारी स----
जाती मोहला १	भार जाती मालवाल १	जाती जाजेदो १	रावगी वाणो १
तेजो पटेल	डुगे पटेल		
वुराको गुजर	जगनाथ		
जाती टकलो १	को बामण		
	जाती वागडो १		
जपाल पटेल	होले पटेल		
पीथी को कुमार	देया को गुजर		
जाती सीरसवो १	जाती डोह १		
सा० हरोमोसो	द०		
चौधरी पटवारी			
सरावगी वाणौ १			
मौ० जसंघ पुरा अमल			मौ० साहीपुरा अमल विहारी
विहारीदास को वसे बीड़			दास जी कौ वसैं बीड़ की
की धरती मौ० संव १६८६			धरती मौ सवत् १६८६ के
को साल जीमी गांव बस्ता			साल जीमी असलि

लगाह थीवी० ३००० वादि		कस्वा थे कोस 2॥
कस्वा तलै लागावी० १०००		पछीम दीसा
		४०००
कसवा का लोग	वीड़ कसवा	रोती जाट लोग वसै
वसै सोहासिल	का मौ आई	
	३००	
वाकी असलि जीमी	चत्रो खोरी	भगवान वीसना
जीमी वीधा	जके पटेल	कै पटेल बामण
२०००	जाती घीटालो	जाती खंडेलवाल
कसवा थेकोस	१	१
आघपछीमदसा		
रेडो पटेल	सा० दामोदर	सा० हरदाल--
गोपाल के	पटवारी सरा-	पटवारी महेस-
जाट जाती चाहर १	वाणी वाणो	री वाएयै १
गो० गोविन्दपुरा अमल	१	मौ० कौसनपुरा अमल
श्री माहाराजा जी कैवसै		श्री माहाराजी कैवसो
सा० बदी दास का करि		झाक के गोलेसवत्

कून वसावै संवत् 1694
कै साल जीमी वीघा कसवा
थे कोस एक प० वाकुणदीसो
वीघा ————
४०००

१६९७ सन १०८४ को
साल जीमी असलि--
वदी दास कारकू
न वसावै कसवा
थेकोस १।। दीस
सगुण बी०
३५००

रोति गुजर लोग वसै
उदा पीथा कौ
पटेल गुजर
कुतन १

नराहण पीया
को पटेल गुजर
जाती कुंतण
१

रोती जाट लोग वसै
जाली पौरा —साहनेत
मा को पटेल सी पटवा
जाट जाती री सरावणी
जाजी है वाएयै

दामोदर प० ————
द०
सरावगी ————
१ ०
मौ० सुखखेडी अमल श्री

मौ० सुखखेड़ी अमल श्री
माहाराजा जी के वस्ये से
वाके गेले सवत् १६६७
सन १०४८ को साली सा०
बंदीदास कारकून वसावे
जमी कसवा थेकोस आध
ने रीतकुण ही सा वी० ———
४०००

असलि
३३००
वहो वी०
१५००

खारड़ावगो
७००
अडाहा
पड़ा रहो
वी०
२०००

रोती गुजर लोग वसै
झावै वाहड़ को
पटेल गुजर जातो
चंड—१

सा हीरो
मीसो पटवा—
री सरावगी १

APPENDIX A-2

Jama'bandī *Mauẓa'* Jo Singh Pur, *Pargana* Malarna
VS 1721/1664

Jama'bandī documents record the actual realization (*ḥāl-ḥāṣil*), the realization being broken down under *māl-o-jihāt* and *sāir jihāt*. The *māl-o-jihāt* is further broken under *māl* and *jihāt*. (See Chapter VI). Crop- sharing (*jinsī*) comes under *māl*. Here we get the total produce under each crop, the share of the *ra'iyat* and *dīwān* (state) which was either one-half (from ordinary cultivators) and two-fifths from the rich peasants (see Chapter IV). An interesting piece of information given in such documents concerns the sale of grain at the spot. In this document the purchaser is the *bania* of *pargana* Dausa. The sale prices are recorded. The details of the heads of expenditure then follow. Finally, we have the balance (*bâqi*) :

Actual realization	—	Rs. 723.09
Expenditure	—	Rs. 723.09
Balance	—	nil

जमा बंदी प्र० मलारणा

संबत १७२१

सन १०७२

मो० जो सांध पुर

अ० नगद

७२३-॥

माल जहाती

६६६ ॥। ≡ ॥

जबती बी १४) १

मु०

१७। =

भुपड़ी वागो

६ । । ।)

तहसीलदार

७≡ ॥

साईर जहाती

२३ =

जमाबंदी प्र० मलारणा

बटाईमण १४४० ॥ ७ ॥

मु०

६८२ ॥—॥

नीडोतरा

२≡

रोजीना

२ ॥ = ॥

कयाली ६८ २॥—॥

मु०

४। —

फसलि खरीफ

७२०। =

माल जहाती

६६६ ।।। ≡ ॥

जबती	वटाईमण	
१७ । =		१४४० ।। ७ ।। मु० रुपया
		६८ २। ।—।।
जबती	बटाई	
वणीबी ६ । १	असली सखालै	फरह
वादी २ बाकी ८.४	मणा २६७७। ।	१०६ ।।
द्र 1।) मु०	वादी हीसा रोती	वीडोतरी
११। ।)	मण १३४६	३३) ३। ।
वग्टी बीघा ४। ।४	वाकी हीसा दिवान	तपदार
	मण	१२। ६। ।

द्र० ।।।)
३ ।।—
मड़वौ ३
द्र० १ मु०
≡)

लवाजमा
२ =
दहनीमी १५ ।)
मु०
।।।०।।।

जरीबानै
१४)१
मु० २१ ।
द्र० १५।।।
मु०
।।—।।

१३३१) ।।)

बाटै आधै
मण २६०५
हीसारौती
१३०२।।
बाकी मण
१३०२।।)

वाटे पचा दुवो
७२।।।
वा रौती
म० ४३।।)
बाकी म०
२६

इजाफे
६३।। ७।।
गंज विचौती
३८।। ७।। २५)

बटाई

बाजरो मण ९५३ । । । ८
बेचो वे. वलुवा फकीरा मोहन
मो० वीछी दौसा कानै द्र० २। ।१
तो. २८ रमजान सन
१०७५ रु०
३७७। । ।-

असली सरवालै | फरह
मण १७८८) ही० | ६५। । ।८
रौती मण ६००) | बीडोतरा
ही. दीवान मण | २२,८
८८८ | तफदार
| ८ । ३

वाटे ग्राधो | वाटै पचादुवौ | इजाफै
१७२८) वादी | ६०) | ३५ । ७
म. ८६४) | वादी म. ३६) | गजै
बाकी | बाकी | १८ । । । ६
८६४ | २४ |

बटाई

जुवारी मण ६२) २
बेचा वे. बलुवा फकीरा
मोहन वीछी दोना का
वाण्या द्र २। ।) तो २८
रमजान
२४। । ।-

ग्रसली सखालौ | फरह |
मण ११६ | ४) २ |
ही रौती ५८) | बीडोतरा | तफदार
हीसा दीवान ५८) | १ । ८ | । ।२
वाटै आधौ | ईजाफे |
| २। २ |
| गजै | वीचैती
वीचैती | १)५ | । । । ७
१६ । ८ | |

तीलमण ७१) ५।।
मण ७०।४ मण।।।१।।
बेचा सा० बेचा रैंती नैं
पीतावर द्र०।।५।।
कसबा कानैं तो १८ रजब
द्र।।५।। १।)
मु०
५१।—।।?
असलि सरवालै
मण १३५।।) वादी
६८) बाकी दीवान
मन ६६।।

वाटै आधै
१२३) वादी
६१।।) वाकी

मोठ मण ३५३।।२
मोठ मण ३५३।।२ वे० वालुवा
फकीरा व मोहन वी की दौसा
कानैं द्र० २) ४ तो २८ रमजान
१६८।=

असली सरवालै मण
६३८) वादी रोती ३१९)
बाकी दीवान मण
३१९
वाटौ आधौ

वाटै पाचदुवा
१२।।) वादी
रौती म. ७।।

६१ । ।   वा० दीवान

५)

फरह   फरह

४ ॥ ५ ॥   ३४ ॥ २

वीडोतरा   तपदार   बीडोतरा   तपदार

१ । । ६ । ।   । । ५   ७ । । ।१   २ । । । ६ । ।

ईजाफे   द०   ईजाफे

२ । ४   २३ । । । ४ । ।

गज   वीचैंती   गज   वीचैंती

१ । ८ । ।   । । ।—। ।   १७) ५   ६ । । ๕ । ।

साहिर जहाती

२० । =। ।

झुपड़ी वागौ   नीडोतरा   २४=)

६ । । ।)   जबती   झुपड़ी का

१७ । =   ६ । । ।)

झुपड़ी   मेटवागौ

1 । ।)   5 ।)   मु०

२ ≡

मेट घास

१ ३ ।

वलवाह द०

१ ०

फसली रबी अज साईर

तहसीलदार रो. । ३७ । । का माल टका

६७ । २५ द्र २५ । मु०

२ । । ≡ ॥

तहसीलदार । ३७ ॥

वाजीव माल टका ११२ । २५

द्र २५। मु० ४ । ।)

रोजीना

२ ॥= ॥

मु० दसतूर सीवाई

२ । मु० ३४ । मु०

=) २ ॥ - ॥

जबती बटाई

१ - । २६ ।

कयाली ६८२॥ – ॥

४। –

द०

मीजालीक

७२३ – ॥

तहवीलाती

७०२ ☰

तों० जसा पोतदार

१७। – ॥

तहवील मनोहरदास पोतदार रु०

६८४। । । ० । ।

वा नगदी ता0 2 रजब सन १०७५

१ ॥ =

वा० वीचैती

६८३॥ =

मा० रजब सन १०७५ तो १९ रु0

१)

मा0 सावन तो. २३

रु०

२९ । —

वा. नगद

१७।= । ।

मा. सफर सन् १०७५

तो. २६ मु.

१ । । ।=

मा. र. आखरी सन १०७५

तो. १

१ । । ।=)

मा. रमजान

७५)

मा. जीलकाद

। = ॥

मुजराई

२० । । ।=। ।

अहीठी अ. फ. खरीफ

वे. भगीरथ वा न टका

मा. र. अवलि

१३ । ।= । ।

तो. १

५ । । =। ।

तो. ४

८

मा. सवाल

५७७ । । ।=। ।

मुकदमी अ. फ. खरीफ

५५२ । । ।= मु.

७०६ । २५

ई. तो. १४ सफर सन तो ३

रजब रो ११५ रो. ३ ।

मु.

४६५)

भगीरथ जसोसाह

२४७। २५ २४७। २५

ई. तो. ८ र. अवल

तो. ३ रजब रो १४१)

द्र. १ । २५ मु.

२११ । २५

वादी खीवडी ३५३ । १२ । ।

बाकी ३५३ । १२ । । द्र. २५ ।

मु. १४ ☰)

बाकी नदारद

०

६ । । ☰ । ।

जबती बाजरौ ८८८)

१७ ।= द्र. ३०८ । । ।

जुवार ५८ ।) तील

द्र. २ । । । मु. ६६ । ।

२१ ६६ । ।

मोठ ३१९)

द्र. २ ।

मु.

१४१ । ।

APPENDIX

A-3

Yāddāshti Ḥaqiqati Hal Bail Zubānī Paṭel Paṭwārī *Pargana* Chatsu *VS* 1723/1666

These documents are in a very bad shape, eaten by white ants. However, I succeeded in getting them microfilmed. These documents throw light mainly on four aspects. First, they give the number of bullocks owned by each peasant. The following table, along with other things, gives the number of cattle per *asāmī* in 11 villages (including *qaṣba* Chatsu), yielding an average of 2.81 bullocks per peasant. The average in the individual villages ranges from 2.11 to 3.53 bullocks per peasant. Secondly, they help us to determine the degree of stratification within the village. There were rich peasants owning large numbers of bullocks (the names of *asāmīs* and number of bullocks per *asāmī* are given) and the poor who possessed hardly any. Thirdly, these documents give a very fair idea of the extent of cultivation giving details of the total areas of a village, the uncultivable area, the cultivable area and then the area under actual cultivation (*jot*) and *parat* (fallow land) based no the quality of soil i.e. *polach* and *banjar* in both the *kharīf* and *rabī*ʻ harvests. Fourthly, castes are recorded against names of individuals.

For convenience of the reader, the Rajasthan text of the document is preceded by a table summarising its contents.

Break-up of total measured area

Pargana Qaṣba Mauẓa	*Year*	*total area*	*% of un-assessed area (Nābūd) out of total area*	*% of cul-tivable area out of total area*	*In'āmī etc*	*Bāqī area*	*% Jot (actual cultiva-ted) out of total area*	*% of Jot out of bāqī (net) area*	*% parat (fallow) out of Total area*	*% parat out of net area*	*Total asāmī*	*Bull-ocks*	*Nos. of bullocks per asāmī*
P. Chatsu	1666	100	14.21	85.79	7.44	78.35	55.55	70.89	22.80	29.11	2448	6200	= 2.53
Q. Chatsu	,,	,,	36.00	64.00	14.62	49.38	24.82	50.27	24.46	49.73	200	422	= 2.11
Kalyanpur	,,	,,	3.84	96.15	—	96.15	76.92	—	19.23	—	18	58	= 3.22
Kerhera	,,	,,	20.00	80.00	2.20	77.80	59.90	77.00	17.90	33.00	69	164	= 2.37
Kisanpura	,,	,,	10.20	89.80	—	89.80	81.63	90.90	8.17	9.10	24	64	= 2.66
Kotkhawada	,,	,,	11.11	88.89	1.83	87.06	78.16	88.11	8.90	11.89	159	382	= 2.40
Kerherakh	,,	,,	9.00	91.00	—	91.00	72.43	79.64	18.57	20.36	28	79	= 2.82
Govindpur	,,	,,	20.00	80.00	18.00	62.00	54.00	87.00	8.00	13.00	15	53	= 3.53
Chandlai	,,	,,	12.64	87.36	20.00	85.36	81.70	95.78	3.66	4.22	165	502	= 3.04
Tutoli		,,	9.62	90.38	0.35	90.03	78.83	87.56	11.20	12.44	68	210	= 3.08
Thikaria Buzurg		,,	16.60	83.40	7.52	75.88	60.25	79.45	15.63	20.55	33	90	= 2.72
Titaria	,,	,,	6.78	93.22	0.22	93.00	70.00	75.19	23.00	24.81	119	349	= 2.93

१

याददासती हकीकती हल बैल जुबानी पटैल पटवारी प्रगना चाटसु वा साल सम्बत् १७२३ सन् १०७४ मुकररा व मैं सरीकती मो. ८२ वादी ईजारा मो. ४

मो. सोमापुर	मो. दादनपुर	मो. स्यामपुर	मो. हरीकिसनपुर
१	१	१	१

बाकी मो. ७७ को मु. रकबौ बीधा ४१४६६८ वादी नाल खाल वा नदी वा ईसर वागौ. बीघा ५८६७४ बाकी लायत जराती बीधा ३२५९९४ वादी जागीरदार सरीकती वा ताबीनाती सगंही कल्याणदास वा ईनामी बगैरह ३०८४३ बाकी बीधा ३२५१५१

जोत साल गुदसता	पड़त वा बीड़
२३०५२२	९४६२९

पड़तगुदस्ता	बीड़
स. १७२२ री	४९००
८९७२९	
सवील हाल	
३१३००	
मु. आसामी २४४८ मु. बौल	
६२००	

असली हाजरी साल
गुदसता आसामी २२४६
बैल ५६००

इजाफो नवा पालती और
प्रगना स्यौ आया वा प्रगना का
पालत्या नवासीर थे बैल कीया वा
नई आसाम्या गाव का नवा बैल कीया

खेती करी ग्रा. (आसामी)
१६६ बैल
६००
नई आसामी गावां की पुराणी आसाम्या
वा और प्रगना की आ गांव का थे
१६६ मु. अधीका
३३० कराया बैल
२७०

कस्वा चाटसु रकबौ ३६२५० वादी तलाव का छापर वगैरह बीधा १३०४७ बाकी लाईक जराती बीधा
२३२०३ वादी ५३०३
इनामी डोहल्या
४३६३

ता. दीवान कल्याणदास जी वर्ग.
६४०
यादसाईमद खाँ हेमराज वर्ग.

का   हमीर दे का

५००   ४४०

ता. दी.

बाकी वीघा १७६००

जोत गुदस्ता	पडत
६०००	८६००

मुकररा आसामी २०० का बैंल हाजरी

४२२

असली साल गुदसता हाजरी आसामी १६३

बैंल   ४०५

माली आसामी ७० बैंल

१२३

मूघर	२	गेगौ	२
हेमा पटैल	२	नाथौ	१
हीरो	१	मोलौ	२
नाथौ पुन्यहो की	२	दमोदरौ	२

रेखौ टीलाको	२	मथुरौ माली	२
सुन्दर पटैल	२	झांझौमाली	१
तोल्हा पटैल	२	चौखौ	१
ताल्हौ	२	झांमौ गांगा कै	
चौखौ	१	मोकलौ	१
होलौ	२	तेलौ माली	१
परमादौ	२	केलौ माली	२
टीलौ	१	दुणौ माली	२
रेखौ मीनाकौ	१		
उदौ माली	२	वालौ	२
कान्हौ	१	भूदेवौ	१
मानौ	१	दुसरो	१
हुगौ रेखौ	१	सालपौ	२
चूहर	२	जैरामौ	२
दीपौ माली	१	डुमौ	२
घरमौ	१	नीवौ	२
भीवौ	२	तुलछौ	२

सीहौ	२
गेगौ	१
रामदासौ	२
गेगौमाली	२
चेतनौ माली	१
भीखौ	२
मूधर	२
मांणमाली	२
रायमलौ	२
भारजौगोरह	२
गुजर रुड़ा कौ	२
तौलौ मारमल को	२
टोडरौ	२
तेसौ	२
चतरौ	२
कुलटौ सांगा कौ	१
नैतौ कुंभाकौ	२
पटेलौरो	१
भगवानौ	४
जगनाथ	४
गणेसौ	२
नरस्सघौ	१
रेखौ पुन्य कौ माली	१

गणेसौ	१
हररामी	२
नरायण	२
नेतौ	१
बिसनौ	१
हैमौ	२

गुजर आसामी ८ बैल

१८

करो	मानौ	नेतो	देईदासो	रायलौ	चलरौ
३	२	१	३	२	४
नाथौ	मुथरौ				
१	२				

तेली आ. २८

५१

तारौ	३	पेमो	१
वेणो अरजन	१	ठाकुरो	१
वेणो सईदरा को	१	टोडो	१

दुदी	१
वेलो	१
वेणो चोखा को	३
तानो	२
परसौ	२
कसेडौ	१
स्यामौ	३
वीरमो	३
रुपसी	१
वेणो केस्याकै	२
वोड्यो	२
लोघौ	१
भरमौ	२
चुहड्यौ	१
माधौ	२
डुग मेहतर	३
केसो	२
हठीलौ	१
ताप्यौ	१
रेखौ	३
पुरो	२
झाझु	४

सम्बत १७२३ सन १०७४

कसबा चाटसु

बाह्मण श्रासामी ५१

११५

ठाकुर सी	कुसैल पटैल	मनौ	वीजौ	हरखौ	
२	२	१	१	२	
जगजीवन	राणौ	कान्हौ	माईदास	करमसी	
३	३	२	२	२	
पुरौ	भावसीध	गोवदौ	सखणौ	घरमौ	
२	२	४	१	२	
दीपौ	जगनाथ	गीरधर	टीलौ	सरीधर	
२	२	२	१	२	
कलवौ	वीणे	केसो	उदै	धरयौ	चलरौ
३	१	३	१	५	१
कचरौ पटैल	घारु	पाथौ	फादु	नरौत्तमौ	
२	४	४	१	१	

दच्चाल खाखबौ	दुरगौ	गाँवदौ	वक्सौ	
३	३	२	२	
जैरामो	केसो	सुदर	वीज	हरराम
२	२	२	४	२
जैतो	लालै	भगपान	मनो	रामदास
१	३	३	२	२
चतरौ	हरराम	दच्चाल पटैल		
५	३	२		
ठाकुर सीगैदौ	चौ० जीणदास	गोरधन चौधरी		
१	४	१		

द० आसामी ३६

बैल ६८

लोधा आसामी ५

१७

हरदासै	डुगौ
६	२
मैंतौ	चतुर

कुमार वा नीलगर ४

११

दमोदरै दुदै	ईसरो कुमार
कुमार ६	१
अलैहदाद नीलगर	पीरसा नीलगर

५	२	२	२
परमानंद			
२			
काईथ आ० ५		रजपुत आ० ३	
११		१२	
भागचंद	बीजैराम	परसोराम जादम	भरहर
३	१	८	२
नराईणदास	मुरारीदास	भीखौ	
३	२	२	
जैकीसन			
२			
	ओसवाली आ० ३		मीणा आ० २
	१०		२
अमीसा	पंचाईण	गोपालौ	खौद्यौ
५	४	१	१
चांदमा रावत			
१			

नागौरी आसामी १४

बैल ३५ द०

ततार	सुल्तान	सबदल
४	४	१
कवरसी	सहवाज	फतु
२	१	१
फतसां	सुजावल	साह
२	२	२
बकसो	जमालखां	लाडखां
५	२	१
सहवाज	पीरसा	
६	२	

ईजाफौ बैल आ० ७

१७

आसामी ७ नवा दा० गांव मैं अधीका

हुई बैल   हुवा ६

११

सीहो बाह्मण   गोही ब्रामण   रेखो आगै १   केसौ अरगै १

२   २   नवी १   नवौ १

गीरवो   गीरधरी   वालो   नामौ आगै १   दुदौ आगै

२   १   १

जगनाथ   दयालौ तंली   नवो १   १ नवौ १

२   २   तेणा आगे १   कनोड्यो आगे १

८०   नवो १   नवो १

APPENDIX

A-4

Dastūr-'amal (*dastūr-al 'amal*) and *'Amal-dastūr* Chautra *kotwālī*, *qaṣba* Phagui 1713.

This document provides us the details of the rate of incidence levied on different people engaged in trade, like *bichhaitīs* (pedlar), local merchants, *ra'iyatī* and *banjārās*. An examination of the commodities brought to the towns throws interesting light on the local trade. The goods brought consisted of two types : first, commodities meant for long distance trade such as indigo, silk and cotton, and secondly, the commodities for local consumption such as cereals.

Different duties were levied upon local merchants, who bought commodities from the villages and sold them either in the same village where they had purchased them or in the villages around, and the *bichhaitī*, who bought the commodities from adjacent places and sold it in the village. The categories of the persons exempted from such levy have also been mentioned in this document.

श्रीराम जी

दसतूर अमल वा अमल दसतुर चौत्रा कोटवाली कसबा फागुही

०	०
गांव कौ व्यौपारी वांव मौ वसत फरोखत करै अर गांव कौ ही व्यौपारी खरीद करै त्याह नै लागै रु० १) पाछै खरीद करि फरोखत मौ वा पालानौ	गांव को व्यौपारी जिनसी बेचो बछाईत को त्याजै लागै रु० १) पाछौ

मापौ	खर्चा	वेचवा वाला गाव कानौ खर्चौ	खरीदकरवा वाला
। १२ ॥	। ६	। ६	बछायतनौ मायौ । २५

गांव में बछाईत आय वसत बेचै गांव कै खरीद करै
त्याह नै लागै उ पाठै बछायत बेचवा वाला गाव
वा वा० लावरानौ
। ३१ ०

गाव मैं महुवा वीकै त्या० का लागै

गाव कै बेचौ बछायत वेचं
अर गाव के हीले गांव का ले तथा
तहि नै रु० १) गांव के बेचे बछायत
१। ले त्याह नै लागौ
रू० १) पाछै
१।

गांव मौ बछाईत व्यौपारी श्राय वस्त लेचै
अर बछायत त्याह नै रु० १ पाछै
बेचवा वाला लेवावाला बछायत
बछायत नै नौ
। ३१ । २५

तमाखू गाव मौ बिके तहि कै लागै
रु० १) पाछै

गांव कै बेचै बछाइत वंचे
गाव कौ लेह अर गांव कनौ
त्याहै नै १) अर गांव कनौ
पाछे बेचै वछायत
१। ले त्याहु नै रु० १)
पाछौ
१।

बछायत बेचै
बछायत ले त्याह बछाय वेचै

— — — — — —

नौ लागौ रु० १) पाछौ २ । खर्चा कोटवाली कोई लागै नहीं	बछाईत ले त्वाहु नौ लागै रु० १) पाछौ २ ।	खर्चा कोटवाली कोई च लागौ ।	द्र० ——————
गाव में छैंडै गाव की बेचै गाव की ले त्याहनै १ । बछाईत बेचै ब छाइत ले त्याह नै लागौ २ ।	बीकौती के लागै बछाईत बैचै गावै को ले गावै कै बेचे बछाईत कै त्याह नै १ ।		खटिक खाल बेचै ती कौ लागै गाव मैं बछाईत गाव कै बेचौ बेचै गांव का ले गाव के लेह तथा गावै की बेचे त्याह कौ बछाईत ले लागै तीहमै रु० १) पाछौ । २५ रु० १) पाछौ ।१२। ।

आव विकौ त्या की
लागति
पीठ की लागति   माथै लाई
    १ ।   बेचै तोका
                  । २५

गाव को वोहरो वारी ला
गावै का पालती नै काढ़ि
दे तहि को लागौ रु० १) पाछै
        । १२ । ।

सीसौ जसद की गाड़ी आई
वुतरो (उतरी) तो त्याह कौ लागै
गाड़ी १ पाछौ

आगै कदे लागै न आयौ छै अब बरस २
            सु लागै
मोहनपुर वगै० गाव ३ कौ हासिल सुदामदी
से थानी चुकै त० जैल गाव

मो० मोहनपुर रामपुर फागुई
    १   १
रामपुर सादी   द०
  १
कयाली खरीद कर वावी लानै
लागै रु० १) पाछै दमीडी २
तीह मघे ऊपर ल माजुरी की पावै
हरबी दमडी नवा की सरकार

	में दमडी दै १।।
गाव का लोह लाकडी सिण सुत वाण बछायत बेचे चौ तथा बछायत वे वै गाव का ले त्याह नै रु० १) पाछौ ।२५	पुखी (पूरबी) विणजारौ आई जिनसी बेचै त्याह नै लागे रु० १) पाछै ।२५
खर्चा कोटवाली कोई लागौ नही	खर्ची कोटवालीं कोई लागै नही
सीगोई	
जिनावर वा गाडै वा छयाली बिकै त्याह लागै रु० १ पाछै १।	गांव में घर घरा की जायेगी धरती वादा वादी वीकौ त्याह नै रु० १) पाछौ ३।
लोह गावै मे गाव का व्यौपारी बेचै अर गाव का ही ले त्या कौ लागे रु० १) पाछै	

। १२ ॥

मिती मागिसर वदि २ सम्वत १७४३ थे सुदामद
दे वाल ले वाल लिया जाहि

गाव में खरवुजा वा सकरकंद हादाघास
हर पानी खाखलै नीरणी गांव में बछायत
गावं मैं बछायत आणि बेचे त्याह नै लागै

खरवुजा सकरकदं		घास वगै का	
सादी का गाडा वीकै		गाडा लागौ	
त्यानै	लौगो	चौखड़ी	बरी
चौ	बी	१ । १२ ।।	। ३१
१ ।	। २५	अ० ई०	अ० ई०
		१ । ।१२॥	। २५ । ६

गाव मैं कलालदार की भठी
करै त्याहनै लागै मु० हरा
७ ।

गांव मै बछायत कपड़े सुतरसमी जरी वाव
बेचो त्याहनै लागौ थान1 पाछौ
। ६

महाजन वा वा ब्राह्मण

रजपूत काईथ भारण भाट

भोजिग का पडी जोगी धोबी

मुसलमान देस वाल मोमल

अतीरु मनै बागदम कोई मिल लागै

नहीं माफ

	हा० हटवाडे		
	हा० दो बटी		हा० सुत
बछाईत	बछाईत	गाव मे गाव	बछाईत ग्राणी
दो बटी खरीद	ढढ जौडा	का ढ्ढ	खरीद क० रौत्य
का रौत्य को	बेंचे व	कई तर वा	कौ लारौ त्याका
लागै रु० २) पाछै	सूत ले	का सुत ले	पाछै । २५
भायौ	त्याह नै		
। ६	लागौ		
———	व्यौपारी दलाल		
	। १२ ।। । १२ ।।		

गावँ में ढेढ़ जौंड़ा
बेची बछायती
ले त्या का
लागै दुवाली
१२ । ।

त्याके
लागती
। १५ ।
छौकुकाड़ी
१
बछाईत
रैंजा वे.चो
त्या रो लागै
रजा १ पाछै
। ६

कसवा का पिनारा
श्राई वीसक करै त्याह
कना उणी वगै उगानै

हा० बंसका
बजाज चमार
खटीक किया
बछाईन त्याहने
लागै
। १२ । ।

रौगर
बछाइत श्राय
नौ लागै सु बोलने
१ पाछो
। २५

कोतवाली की लागती
रु० ४ की तौ माफ श्ररु

कोतवाली की लागती
जिनसी विर्क त्या सीगोई

रु० २ लार टको एक लीज्यौ	की रु० २) लार	रु० ५) उपर की होय
१।	।२५	ताई तौ को
		।१२॥।२५

अतनी जिनसी की कोटवाली
माफ
तमाखू महुवा सिण
लोट वगै सुखी
बीणजारा

भोमि

जिनसी का मापा	धरती का वाटा
वगै का टका मेला	वादी गाड़ा वाहणी
करि अर लीनो टका	बीकौ त्याह को टका
१।पाछौ	१। पाछौ कोड़ी ८

कोड़ी ८ अर वट
वाली का टका की
भोमनि लागौ

सीगोही रासि १	बाग दम १
पाछौ	पाछौ
।२५	१।

भवसा सगल नै हासिल लेना देना
दरोबस्त माफ

प्रोहत बिजैराम चत्रां वावक
सुवोर रकुल सेवात्वागै डी हलत
नै हासिल लेता व देता सुदामद
माफ छै दरोबसत
पटौल नै माफ

पोहकर पहलाद पटौदा व हीरा का
कौ हासिल लेता देता दरोबसन
माफ झौरु पटैला नै सु दामदी कोट
वाली माफ

हरदा पुरवा का पटौल नौ सा० अखै राज
का अमल थे माफ हुवौ
गुजराती माधै गोरधन का बेटा
त्यानै हासिल समादि मु० प्रवानौ
श्री कै लेना देना माफ

मापाति हासिल मघे वा चौत्रा कोटवाली कस्बा फागई
कासली वाला नै बन्दुक ?
राज० पहाड्या नै हाहिल ले
तदिताद दरोबसत माफ
संगही सोभाचंद बुरमने
हासिल लेना देना दरोबसत
माफ

वलुरेखा का महैसर महाजन नै हासिल
लेना देना दरावसत माफ

जाट पटौला कै घर रोईति की बाग
द माफ लागैनहीं

APPENDIX

A-5

Dastūr-al 'Amal Pargana Udai, VS 1722/1715

This document gives the tax on each crop that the peasants of different categories had to pay. It also gives the details of the demand on the basis of quality of land, i.e. *polach* and *banjar*, and the information of different rates of *māl* (tax on each crop) and *jihāt* (other cesses) is offered under two distinct heads, viz, *ẓabṭī* and *jinsi* on the same pattern as in the *arhsattas*. The rates under the two categories are, strictly speaking, not comparable with each other. The crops put under *ẓabṭī* were assessed in terms of money per *bīgha*. The *jarib* (measurement) used in *pargana* Udai was *gaz-i-Sikandarī* of 60 yards and the length of the gaz was 38½ *anguls*. Different rates were fixed on *ra'īyat* and *qānūngo*, the details of which are as follows :

Rate per *bīgha*

Crop	(in rupees)	*Ra'iyat*	*Qānūngo*
Cotton (*van*)		2.00	1.50
Sugarcane (*varh*)		4.87	4.10
Tobacco		4.00	3.00
Jute (*Sunn*)		2.37	1.87
Barti		0.75	0.50

Jinsī (*Bhaoli-baṭā'ī*)

Quality of land	*Rate*	*Share of State*	
Polach		50%	*Ra'īyat*
Banjar	Two-fifth	40%	*Ra'iyat*
		25%	*qānūngo*
		40%	*mahājan*
		33%	*Rajputs*
		30%	*Kheldār etc*
			paṭel
		40%	*paṭel pālti*

The *qānūngo*, Brahman, *mahājan*, *patel* and *kamīn* were exempted from the *jihāt*. The *ra'īyat* had to pay *jihāt* at the rate of 3% of the amount of *māl*.

दस्तुर अमल प्रगना उदई सम्बत १७७२ (?)
दस्तूर अमल प्रगना उदैही सरिकार सूबै मुसलतसल
खीलाफती अकबराबाद
फ० खरीफ

माल
जव्ती जरीब गंज सोकदरी आगूल ३८ । । डौरी गंज
६०

वण बीघा (१) पाछै
रैती बी (१) कानूमोवा हीरामन
पाछै भीवा बगस र्री का
(२)
बीघा 1 पाछै
१ । ।

रैती
सीचवार
(बी १)
पाछै
४ । । । =

वाड़ बीघा (१) पाछै
मारुपडी
(बी १)
पाछै
३ । । ।
सचिवा
बीघा
(४)

कानूनगोह
हीरामन
भीवा बससरी
(बी १)
पाछै
भारपड़ी
बी १
पाछे
(३)

तमाखू
रैती
(४)

सन
कानूगो
(३)

पटसन
रैती (२ ।=)
कानूगो १ । । =

वरटीकुरी
रैती (। । ।)
कानूगो (। ।)

भावली बटाई

रैती पोलच जमी आधो

रैती दी०

१ १

जमी बंजर पच दुवौ

रैती दी

३ २

कानुगो की वाटो नीखरचो चोथो

४

रैती दीवाना

३ १

हीरामन भीवा बकस री बाकी
वाटो तीसरो

३

रैती दीवान

२ १

माहाजन के वाट पच दुवै

रैत दीवान

३ २

ब्राह्मण भाट कस्वा को वाटो
आधो नीखरचो

२

रैती दीवान

१ १

राजपूत को वाटै

राजपूत के कौटरी के
वाटे तीसरो वाटे चोथ

३ ४

रै० दी रै० दी०

२ १ ३ १

माहाजन कसबा को वाटो
नी खरचो लीजौ प्रगना का
माहाजन की वाटैं खरच प्रगना
वरबसती लीजौ

खेल दार वगै पटैल
कस्बा का आसाती 12
को वाटो तीसरो

३

रैत दोवान

२ १

पटैल पाही को वाटै
पचै दुवै तै जैल
रैत दीवा

३ २

खरीफ

खैलदार आसामी

८

ईसमाहल रु० मोहम्मद

१ १

नीवाज खां ततार

१ १

मीठो खां याफुच

१ १

पाही कौ तो वाहो हमेशा
पचदवै वटै छै अर पटैंला
प्रगनौ का संबत 1771 की
मैं हूजुर उजन गया छा
सो माहाराजा श्री राज करैला वो वाह
वाने पधरा छै तब रैत सारी लार
आई सो कस्बा उदैही का डोर

हाजी हैं खां

१ १

गोला पटेल

आसामी

४

जुलाहे घूरा

१ १

हीरानंद चतरु

१ १

राम राई सूखा हीरजामन

दयाराम ब्राह्मण का वाट

पच दूवै

५

रैत दीवान

३ २

ओर चोली राम राजाराम चौतराम

हीरदराम आसामी ४ बाह्मण का वाटो

अमल माहाराज का सू वाटै पच दूवै

कमीनी कीसान थे मागले तीको वाटौ

तीसरो नीखरचो

५ ६ वाम्हे दीलासा कर हुकम

हुवौं पटैल तटैलासू अमल

सगं ही जग जीवन दास साहीबराम

का अमल पचदुवो पटैल

वा पाही को वाट पच दुवो

सीग हर्या को वाटै

पच दूवै

रैत दीवान

३ २

रैती दीवान
२ १

जीहाती

जीहाती वा नकदी वा
ईजारा की जमा चुकै
ती पे लागे द्र. सैकड़ा १००)
पाछै ३)

भुसी जीनसी सीरकार
हीसापे लागै मन १)
पाछैदाय पचीस
।२५
कानूगोह का हीरामन
भीवा बगस री का ब्राह्मण का भाट
वा माहाजन वा पटैल कसबाका रंजत
वा कमीन माफ

रबी

माल

जबती जरीब सरैह संदर

पोसत अजवानीं वा
कुसुम बीघा १) पाछै
४)
काकड़ी का साग लाल्हरा
बीघा १) पाछै
१)

लहसुनी वा बगन बीघा १)
पाछै २)

कादा बीघा १) पाछ
२।।)

मीणा बीघा १) पाछ
रैती बी० १)
पाछै १ । ।)

कानूगोह वा
हीरामन मीना
बगस द्र० वी० १) पाछै
१)

भावली

रैती को वगै
जीमी पोलच को जीमी वंजड़
वाटो आधो वाद फसली
२
रैती दी वा सेका कौ
१ १ वाटो पच दुवौ
५

रैती दीवान
३ २

हीरामन व भीवा बक्सी रौ वाको
वाटो तीसरो
३

कानूगोह कौ वाटो चोथौ
नीखरचो
४
रैती दीवान
३ १

ब्राह्मण व भाट को वाटो
आया नीखरचो
२

रैती दीवान
२ १

महाजनो को वाटो पचदुवो
५
रैती दीवान
३ २
महाजन कसबा को वाटो
नीखरचो लीजै अ्रर प्रगना को
महाजना को वाटो बरबसत प्रगना
के लैणो
पटैल वा पाहीश्रान प्रगना
का को वाटो पच दूवो
५
रैती दीवान
३ २
सम्बत १७७१ की स्यालू सु
वाटो पचदुवो माफक सरहै
सदर

राम राय सुख व हरनाम व दयाराम

रैती दीवान
१ १
सेको का वाटो पच दुवो
बेलदार व गौला पटेल
कस्बा का आमामी १२
को वाटो तीसरो
३
रैती दीवान
२ १
आसामी १२ को माफक
सरहे सदर
राजपूत को वाटो मु० तपसील
जैल नीखरचो
रजपूता को वाटो कोटरी को
तीसरी वाटो चो थो
३ ४
रैती दीवान रैती दी०
२ १ ३ १

कमीनी की साना कनै

को ब्राह्मण को वाटो पच दुवौ		भाग लेती को वारो	
नीखरचो ५			
रैंती	दीवान	रैंती	दी०
३	२	२	१

चौली राम राजाराम चेब राम
हठीराम ब्राह्मण की वाटो श्रमल
महाराजा का सु पच दुवो

जीहाती

जब्ती व नकदी व
इजारो चुकती शेर
लागै द्र० सैंकड़ा पाछै
१००) मु०
३

भूसी हीसै सीरकार का
मण १) पाकै दाम
पचीस
२५

कानूगोह व हीरामन
मीवा बक्स री या ब्राह्मण
माट ब महाजन व पटैल
आसमी १२ कसबा
सब राजपूत कमीनी माफ

APPENDIX

A-6

Taqsīm pandrahsāla, pargana Antela Bhabhra, *sarkār* Alwar, *ṣūba* Akbarabad, mauẓa 17, *'aṣlī* 11, *dākhlī* 6.

Total *bīgha*	61,110.0 .0
vadi sir, pahar, nadi, nāla	21,380.0 .0
Baqi *layaq zarā'it*	39,800.0 .0

Year	*VS/AD*	*Total Revenue Realized* (in rupees)	*Harvest-wise Break-up* (in rupees)	
			Kharīf	*Rabī'*
H. 1057 *VS*	1706/1649	9870. 5.0	582. 2.0	9288. 3.0
H. 1058 *VS*	1707/1650	15383. 3.0	8403. 0.0	6980. 0.0
H. 1059 *VS*	1708/1651	13386. 8.0	7513. 0.0	5873. 8.0
H. 1060 *VS*	1709/1652	13784. 0.0	8801. 8.0	4982. 8.0
H. 1061 *VS*	1710/1653	6922. 0.0	989. 0.0	5933. 0.0
H. 1062 *VS*	1711/1654	14470. 8.0	8841. 8.0	5629. 0.0
H. 1063 *VS*	1712/1655	12029. 4.0	5799. 4.0	6230. 0.0
H. 1064 *VS*	1713/1656	10801. 3.0	5913. 4.0	4887.15.0
H. 1065 *VS*	1714/1657	12549. 8.0	6426. 0 0	6123. 8.0
H. 1066 *VS*	1715/1658	13943.12.0	3086. 0.0	10875.15.0
H. 1067 *VS*	1716/1659	11501. 0.0	10146. 0.0	8355. 0.0
H. 1068 *VS*	1717/1660	16987. 0.0	7016. 0.0	9881. 0.0
H. 1069 *VS*	1718/1661	13800. 4.0	7299. 8.0	6500.12.0
H. 1070 *VS*	1719/1662	13151.12.0	6214. 4.0	6937. 8.0
H. 1071 *VS*	1720/1663	10781. 4.0	894.11.0	9886. 9.0

These *taqsīm* papers were prepared by the *qānūngos*.

Taqsīm dahsāla (ten years), *pargana* Udai (1734-1743)

The statistical records of the measured area and the assessment made upon culturable land of a *pargana* (including villages) are contained in *taqsīm* and *muwāzana-i-dahsāla* documents.[1] The length of the *dori* used in Eastern Rajasthan consisted of either 75 *hāth* (i.e. *kachhidori* or 95 *hāth* (*pakki dorī*). In *pargana* Udai, a *dori* of 95 *hāth* was used. However, the yard used in Udai was the *gaz-i-Sikandarī*.

These documents show the proportion of cultivable land (*lāyaq zarā'iti*) out of the total measured area (*kul raqba*), which also included *nābūd* (uncultivated) and so remained unassessed. The *taqsim* of *pargana* Udai provides us the following details :

	Bīgha	*Biswa*
Total area (*kul-raqba*)	3,82,542	16
Nābūd	50,339	7
Cultivable area	3,32,203	9
Inhabited (*bastī*)	2,655	10
Thola	2,247	10
Wells	492	12
Trees	135	00
Nāla	4,857	00
Tālāb	1,429	4
Bahar (forest ?)	100	00
Path (*rāh*)	1,955	00
River	1,200	00
Hills (*pahār*)	2,240	00
Sarekhar (?)	18,949	00

Such documents may help us to form a rough estimate of the *jama'* incidence per *bīgha* by dividing the estimated *jama'* by the net assessed area.[2]

1. See my article, 'Agrarian Information in *Taqsīm* documents from Eastern Rajasthan (1649-1767)', *PIHC*, Vol. XI, 1979. Jaipur, pp. 57-64.
2. *Ibid*. (Appendix II).

तकसीम दहसाला प्रगना उदैही सरिकार सूबै अकबराबाद जमा मौजा १५३ वांदी वैरान सनवाती

गाव २

मौ० परिखडैहरी	सुलतानपुर
१	१

बादी मौजा १५१ वादी ग्रसली गावां मैं दोखली मौ० २०

बाकी मौ० १३६

मुकररा रकबौ डौरी गजसीकरी हाथ ६५ थै बीघा ३८२५४२)।।।१

वादी नाबूद बीघा ५०३३६)।२

गाँव की बसती बीघा	थला बीघा	कुवा बीघा	दरखत	नला
२६५५)।।	२२४७)।।	४६२)।।२	१३५)	४८५७)
तलाब	बहड़	राह	नदी	सारेरवार
१४२६)४	१००१	१६५५)	१२००	१८६४६)
पहाड़				
२२४०)				

बाकी लायक जराईनी बीघा ३३२२०३)।४ मुकररा दाम

१०००८०००

वादी बाढ़बार सरिकर कानै धरती मापी दीन्ही जैरी हाथ ७५ की थै बीघा मु० नुसखें मुकररा डौरी सीकें हाथ ६५ की थै बीघा ?? २३५) ? बाकी बीघा ? (?) मुकररा जमा तफसील जैल० अमल खालसा, जागीर वा ईजरा सुधा

दसखत कानगोह सवाई जयसिंह नाहरमल

प्रगना उदैंही

	हासिल	खरीफ	रबी
सम्बत १७९१	९९९३६	४९४८२	४७४५४
सम्बत १७९२	१०७५८१)।=	३८९९५)।।=	६८६१६)=।।
सम्बत १७९३	८५४८३	४६०४१)।।—	३८४४।)।☰
सम्बस १७९४	१०००१०)।—	५०७४०	४९२७०)।—
सम्बत १७९५	११०५९९)।।।—।।	६१२०९)।।।—	४९३९०)।।
सम्बत १७९६	१२१०५८)।=	७०१७४।।—	५०८८३)।=
सम्बत १७९७	१३५१४८)।।	७४०९४ ।—।।	६१०५४)☰।।
सम्बत १७९८	१०६५०६)।—	५८७०६—	४७८००)।।।
सम्बत १७९९	१००००२)	५१४७५	४८५२७
सम्बत १८००	१०८८२०)।।☰	५३१९)।☰	५५६२९)।☰

तकसीम उदैंही

मौ० कमरी बुजक मो० असली १ दाखली मौ. स्याहपुर १

मुकररा रकबा बीघा ८०४६) वादी नाबुद ७००)

गाव की बसतीं	सोरखार	नाल	तलाब	राह	दरखत
४०	४००	५०)	१००	५०	५०

कुवा

१०

बाकी बीघा ७३४६ लहिक जरहित मु. जमा वादि बाढ़ डोरी हाथ ७५ की थे २३२०) मु. पकी डोरी को १४५०) बाकी ५८६६

मु० जमा	जमा	खरीफ	रबी
सम्बत १७९१	२६२५)।।	१२५१)	१३७४)।
मम्बत १७९२	२९६०)।	१४१४	१५४६
सम्बत १७९३	३७००)।—	१७६७)।।	१९३२)।।।–
सम्बत १७९४	३९३५)।=	१८२१	२११४)।=।।
सम्बत १७९५	४०११	२००५)।।	२००५)।।
सम्बत १७९६	४२२३	२११२)	२१११)
सम्बत १७९७	४२२३	२११२	२१११)
सम्बत १७९८	४२२३	२११२	२१११)
सम्बत १७९९	५५००	२७५०	२७५०
सम्बत १८००	५५००	२७५०	२७५०

APPENDIX

A-7

Khasra of Village Morara, *Pargana* Naraina VS 1802/1745

A large number of *khasra* papers have survived from Eastern Rajasthan, particularly the Jaipur and Kota region. In Kota, *khasras* are found in a series. Similar, documents like *chuthār*, *khāta* and *tazkara* are equally important as *khasra*. References to *khasra* occur frequently in *arhsattas*. These village papers were usually maintained and duly signed in the local language by the *patel* and *paṭwārī*, the headman and the accountant respectively. The *khasra* papers are of different kinds, viz. *khasra māpā*, *khasra jama'bandī*, *khasra kūnt* (*kūt*), *khasra batāī*, *khasra lātā*, *all* giving a variety of information. *Khasra* documents contain in brief the following entries :[1]

1. Name of the *asāmī* or cultivator.
2. Width (*'arẓ*) of the field of the *asāmī* under each crop.
3. Length (*ṭūl*) of the field of the *asāmī*.
4. Total area, calculated by multiplying No. 2 by 3.
5. *Vadi* or exempted area on account of *nābūd* (crop failure, *kharāba*).
6. *Bāqī*, the balance of the cultivated area, after deducting No. 5 from No. 4.
7. Cash rates per *bīgha* (in *khasra* documents other than these given here) applied for the respective crops. Entries Nos. 2 to 7 are furnished for *ẓabṭī* crops only.
8. For other crops, on which revenue was assessed in kind. Total produce (*jumlo*) of crop cultivated by each *asāmī* and the share of *dīwān* (state) and the *ra'iyatī* (cultivator) out of it.
9. Other exactions levied upon peasants under *ẓabṭī* and *jinsī* along with their rates.
10. An abstract account completed from other detailed accounts.
11. The details of the various rates of taxation in kind

1. S.P. Gupta, '*Khasra* Documents In Rajasthan' (See Bibliography).

(*jinsī*) upon the peasant such as *adho* (50%), *tisro* (33%), *pachduvo* (40%), etc.

Besides the crop and the area sown by each cultivator, the *khasra* document tells us about the individual cultivator's holdings, which helps us to determine the extent of stratification among the peasants.

These documents have some limitations. The area under *ẓabṭī* does not give any indication of either the produce per *bīgha* or the value of the crops. But in the case of the crops, assessed according to the *jinsī* system, the total produce is given along with the quantity demanded as revenue.

Naturally, the generalisation one can hazard on the basis of information relating to single villages can only be reliable in a limited way. But the studies of these valuable documents over a number of years may still help us to fill up many gaps in our knowledge of the agrarian economy.

श्री रामजी

खसरो मौ. मौरड़ा प्र. नरायण वा. साख स्यालु सम्बत १८०२ कोला टोडी रीवाणी तीको मिती फागुण सुदी ८ सोम्वार

जीनसी कुल

आसामी	जुवारी	बाजरो	मोठ	मूंग	तिल
बकसी गौरो	०	।५	०	०	०
ग्रादमपलि					
तालीक पटैल					
सुन्दर को	०	।।)	०		
तालीक पटैल देवो	।५५। सौ	।।।) देवो			
गौधोपाल					।।)
जीवो गुजर	।।।)	।।।)			
दैवो पटैल	१)	१।।।)			
नुपो महाजन	०	।।)			
हरजी जाट	५				

नुरो पटैल	२	०		०	०
जौधो रपाडे	०	।।)	०	०	०
लिख मैणु	०	।)			
ईसर	।५				
आसो जाट	।।)			।)	
कैसरो पाजै	०	।।)	०	।।)	
दैवी संघ ठाकुर					
प्र० पान नाकै	१०)				
औसकरण		०	५	०	०
	१५।।५	५।।।५	५	।।।)	।।)

मुली

२२।।।५

जुवती जुवार

आसामी	अरज	तुल	वीघा	गाजर	सकट
खेमो पटवारी	५१	४६	५।।२	२	१।६
लाखो गुजर	५४	३२	४।१	०	४
ईसर बाड़ा वालै	५०	४।	५।।।	—	०

जानसंघ राडौ	५६	२३	३।२	१	।)
खैतौ गुजर	३५	३०	२।।३	०	०
नानीग रपाडै	२८	२६	१।।।१	—	—
परसो रपौडै	५५	३३	४।।१	०	२
ठाकुर सदाराम					
वा गंगा लाखा	३४	२८	२।३	०	।२
कान्हो बैण					१
नया रादा	३५	३१	२।।४		
आदम पटैल	३६	२७	२।४	२	४
कानो कुम्हार	२८	३२	२।	०	१।१
भौमरो संघ			०	०	१
नाथो	१५	१५	०	०	।।१
अजुनो	०	०	०	—	३
ईसेरो	५२	३५	४।।२	०	—
आसो बरड़	५५	२७	३।।४	०	१

॥ श्रीरामजी ॥

गीणाती

आसामी	मैंसी	झोटी	छैली जाय
खेतो गुजर			१६३
बुध स्पाखै	१	१	१५
जोधो पटैल	३	४	७६
चैनो गुजर	३	—	०
जीवो छौकर			४७
बीन स्पसंध			४
लीखमो गुजर	१	१	१०८
खीतो गुजर			१४६
कौजु जुलाहा		०	२
परमी संघ			२
अमरो राजपूत	१	१	०
हरजी बारठ	१	०	३
अजब संघ चंदी	३	१	०

चदो जोधी	२	२	०
लिखमो जाट	१	१	१४
परसो कुम्हार	१	—	—
सुखो तैंतर जाट	१	१	०
परसो रजाड़	१	०	०
लिखमो तैंतर	१	१	—
सुओ जाट	१	१	—
दैवो पटेल	२	१	—
सदाराम खगोव	२	२	२५
लखत संघ	१	१	०
हरजी बरड़	१	२	२
कूसलो जाट	४	३	२
ईसर बरड़ जाट	३	२	—
रूपो पटेल	२	१	१८
लेखो बरड़	०	—	२
दोलो सारण	१	१	०
ओसो बरड़			३
दोलो सारण	१	१	०
आसो बरड़	०	०	३
केसरी संघ	०	०	७

नैनो जाट	०	—	१३
मनो गुजर	—	—	१५२
रूपो गुजर	—	—	२६६
नाथूई०	१	१	—
आदम पटैल	०	१	—
	३८	२९	१०७४
	रैत ह०	रैत ह०	रैत ह०
	२१ १७	१४ १५	९२४ १५

दुजडो

जुरौ जुलाहा	कौजियो तीनाणा	अणदो बलही
१	१	१
जीयो बलह	खरयो बलह	अखलो
१	१	१
घेनो सुधार	द०	
१	०	

लिखतु मौ० मोरढ़ा का पटैल पटवारी अत्र सम्बत १८०२ फागणी तीज फड़ी आग होई कराई बीन इपतफायुत होइत मदै जाब करा मिती फागण बदी ८ सम्बत १८०२ का

खौर भैस	इनमासु जोटौ	इनमासु गाय
३८	२९	१०७४

इनामासुथ

।। श्रीरामजी ।।

आसौ बरड़ जाट	५४	२८	३ ।।। १	०	—
लालो बरड़ जाट	१४	१३	। ४	—	)२
अजनौ	१४	३०	१)१		
केसरो पटैल	१	६	०	०	। १
नैतौ बरड़ जाट	४०	३५	३।।	०	—
कैसरो पटैल	३५	३४	२ ।।। ३	२	०
कुसलौ बालबाड़	४५	४८	५ । ३	२	०
लिखो बरड़ जाट	३४	३२	२ ।।।	—	३
हरजी बरड़ जाट	२७	३२	२) ३	०	०
उटका			२		
हरजी को जुमो	१६	२८	१) २	०	०
रूपो पटैल	१६	१३	।।)	०	०
अज़नो	२६	१६	१ । ४	१	—
देवो पटैल	२८	२६	२)1	०	०
रूपो पटैल	१८	१७	।।।)		
अजनो	२८	३१	२)३	२	४
हरगी कूजुडो	४८	२६	३)२	०	२
दूगा			०	—	३
जोधो पटेल	५५	२७	३। ।४	४)२	०

सदाराम खगावत	४६		३६	४)३	३
अजबसंघ चंदी	३०	३६	२।।४	—	—
बुधसंव	४८	४३	५)३	२	३
धुड़ जागु	३२	३५	२।।।	२	०
जौधो रपाड़	४७	३३	३।।।३	०	४
नुरो रपाड़	२९	३४	२।४	०	०
लखो वरड़ जाट	२७	५२	३।।	०	२
श्योप संघ	३४	३३	२।।।१	०	३
लीखमो मोतीवरड़					२
देवो पटैल	४८	२४	२।।।३	२	४
पीस्ली संघ	३५	२८	२।४	३	३
			११६)३	११४	४।।।४

बासे १)४ बादो १

११४।।)४ बकी ४।।।३

तै रोज पुजी :

मण मैण

आसामी	मण
जुवारो	१५।।५
बाजरो	५।।।५
मोठ	५

जीवती बीघा

आसामी	बीघा
जुवारी	११४।।।४
सबजी गुड़	४।।३

मूंग ।।।)
तिलैं ।।)

——————

२२।।।५

कानूगोह प्र. नरायणा को मुड़संघ सम्वत
१८०२ को जोटोडी रो जमा आज र
जवती बी जिनसी सरा
११६।।।२ जीकणा
२२।।।५

११६।।।३
। लिखतु मौ. मौरड़ा
का पटेल पटवारी अत्र
पटवारी अत्र
साख सालु सम्वत १८०२ का अग
होई करायैं दीवां हमत फावत होईत
जाब करावो जनस वखत को जनसी
जबती २२।।। ५
११६।।।२

गौहू वाट तीसरै

आसामी	सरवाल	रैती	दीवान	सैरी	जुमला
हठीराम के बीघा	३६)	२६)	१३)	।।।२।।	१३ ।।।२।।
२८) ७) ४)					
नीरसंघ वा.	४८)	३२)	१६)	॥ ८	१६।।८
३५) ८।।।) ४)					
सदाराम कुं.	७।।)	५)	२।।)	४।।)	२।।४।।
७।।) १।।) ।।)					
रामदास बाडै	६)	४)	२)	३।।)	२) ३।।
३।।।) १) १।)					
मधुर पंचोली	६७।।)	४५)	२२।।)	१)	२३।।)
४०।।) १०।।) १६।।)					

गौधो दीवाण	६।।।)	४।।)	२।)	४।।)	२।४।।
४।।) 2।)					
वीण पटैल	२५।।)	१७)	८।।)	।५	८।।।५
१६।।) ४) ५)					
खेतो वा. गंगाराम	७२)	४८)	२४)	१)	२५)
४५) ११) १६)					
मयाराम गंगाराम	६)	४)	२)	३।।।)	२)३।।
३।।।) ।।) २)					
लाखो पँचोली	५७)	३८)	१६)	।।।४	१६।।।४
३३।।।) ८।) १५)					
परसराम रावत	२४)	१६)	८)	१४	८।४
१५।।।) ४) ४।)					
रैंवड़ो पंचौली	१८।।।)	१२।।)	६।)	।)	६।।)
१२) ३) ३।।।)					
जगरूप पटैल	३१।।)	२६)	१०।।)	।६	१०।।।
२२।।) ५।।) ३।।)					
खैतसी	२७।।।)	१८।।)	६।)	।।)	८।।।)
२०) ५) २।।।)					
छत्तर पटैल	५१)	३४)	१७)	।।।)	१७।।।
३२) ८) ११)					

दलपत्ती	८७)	५८)	२६)	१ । १	३०। १
५४।।।) १४) १८।)					
ईसर नाल नोल	१६ ।।)	१३)	६ ।।)	। २	६ ।।। २
१२(३) ४ ।।)					
लालो पटैल	३०)	२०)	१०)	। ८	१० । ८
१६ ।।) ४ ।।) ६)					
पैंमु पटेल	१३ ।।)	६)	४ ।।)	)८	४ ।। ८
६ ।।।) १ ।।।) ५)					
मवाराम जोवु	२१)	१४)	७)	८ ।।।)	७)८ ।।।
६) २ ।) ६ ।।।)					
पटवारी का काचो					
छैराका	६)	४)	२)	०	२)
	——	——	——	——	——
			२२१ ।।।)	६ ।।। ६।	२३१ ।। ६।

गोहु पचा दुवा

आसामी	जुमला	रैंती	दीवा	सैंरी	जुमला
हीरो बंडलो	२१ ।)	१२ ।।।)	८ ।।)	। ५ ।	८ ।।। ५ ।
१५) ४) २ ।)					
हरराम वा.	२५)	१५)	१०)	। ८	१० । ८

१७।।) ४।।) ३)

घानू बामण	२४)	१४) ६	८।।४	।८	१०) २

१७।।) ४।।) २) ३)

नीरायणा मीरधा	२२।।।५	१४)५	८।।।)	।६	६)६

१५) ४) ३।।।५

।। श्री रामजी ।।

वाट साढ़ तीसरैं

आसामी	जुमल	रैंती	दीवान	सैंरी	जुमल
मुधर पचौली	१०६)	७५ ।। ८	३० । २	१ । ३	३१ ।।५
७०) १८) १८)					
लाखो पंचौली	१७ ।।)	१२ ।।)	५)	६)	५)६
१२ ।) ३) २ ।)					
मौला काईथ	६६ ।।।)	७१ ।)	२८ ।।)	१ ।)	२६ ।।।)
७३) १८) ८ ।।।)					
			६३ ।।।	२ ।।२	६६ ।। ४

तैंरीज कुली जौह बैंभी गौजरो

आसमी	वट	सैंरी	जुमल
वाट तीसरैं	७३१।७।।	३१। ।६। ।।	७६३)४।
वाट पचा दुव	३६२। ।६	१५। ।। ६ ।।	३७८ ।। ५ ।।
वाट चौथ	२७८ । ० ।।	१२ ।) ।।	२६० ।। ७
वाट साढ़तीसरैं	६३) ।। २	२ ।।। २	६६ ।। ४
	१४३६ । २	६२ ।। ८ ।।	१४६६) ।।।

।। श्री रामजी ।।

कमीण			
आसमी	जुमलैं	रैंती	दीवान
दलाराम नाई	। ५	।)	५
चंदो खात्ती	९)	६)	३)
जुहर	९)	६)	३)
हरनामू नाई	३)	२)	१)
दीपो कुम्हार	।।।)	।।)	।)
गोधो नाई	। ५	।)	५
गंगलो नाई	। ५	।)	५
रैंवडी नाई	। ५	।)	५
कुसलो चाकर	।।।)	।।)	।)
बलाही	३।।।)	२।।)	१।)
	२८।।)	१९)	९।।)

APPENDIX B-I

Silver Value of Gold (rupee per *muhr*)

Year	*Amber/ S. Jaipur*	*Bahatri*	*Chatsu*	*Dausa*	*Lalsot*	*Malarna*	*Niwai*	*Pirayan*	*Pahari*	*Toda Bhim*	*Udai*	*Wazirpur*
1	2	3	4	5	6	7	8	9	10	11	12	13
1665	—	—	15.12	—	—	—	—	—	—	—	—	—
1666	—	—	—	—	—	—	15.00	—	—	—	15.00	—
1688	12.00	—	—	—	—	—	—	—	—	—	—	—
1690	—	12.25	—	—	—	—	—	—	12.25	—	—	—
1693	—	—	—	—	—	—	—	—	—	—	15.00	—
1694	—	—	—	—	—	—	—	12.31	12.31	—	—	—
1695	—	—	—	—	—	—	—	—	—	—	—	—
1702	—	—	—	—	—	—	—	—	—	11.06	—	—
1703	—	—	—	—	—	—	—	—	—	11.25	—	—
1712	—	—	—	—	—	—	—	—	—	11.50	—	—
1713	11.75	—	—	—	—	—	—	—	—	11.25	—	—
1714	—	11.87	—	—	11.75	—	—	—	—	—	—	—
1715	—	—	—	—	—	—	—	—	—	—	—	—
1716	—	12.00	—	—	—	—	11.75	—	—	—	12.50	—
1719	11.75	—	—	—	—	—	—	—	—	—	—	—
1720	12.00	—	—	—	—	—	—	—	—	12.00	11.75	—
1721	11.87	11.75	—	11.62	11.75	—	—	—	—	11.69	—	11.94
1722	11.75	—	—	—	11.75	—	11.75	11.53	11.53	11.69	—	11.94

1723	12.00	—	—	11.62	11.62	—	11.62	11.75	11.75	11.75	11.75	11.81
1724	11.75	—	—	—	—	—	—	—	—	—	11.75	—
1725	12.06	—	11.87	—	—	—	—	12.75	12.75	—	11.75	11.87
1726	—	12.75	—	—	—	—	12.00	12.00	12.00	—	—	—
1727	12.19	—	—	—	—	—	—	—	12.00	12.75	12.06	—
1728	12.19	12.06	—	—	—	—	—	—	—	—	12.25	—
1729	12.50	—	—	—	—	—	—	12.50	—	—	—	12.25
1730	12.50	—	—	12.62	—	—	—	—	—	—	12.25	—
1731	12.62	—	—	—	—	12.50	—	—	—	—	—	12.44
1732	—	—	—	—	—	—	—	—	—	12.44	13.00	—
1733	12.62	—	—	{12.50 {12.37	—	—	—	—	—	12.44	—	—
1734	11.87	—	—	{12.75 {12.87	—	1275	—	—	12.50	—	—	—
1735	12.78	—	—	12.75	—	—	—	—	11.87	—	12.87	12.75
1736	12.67	—	—	—	—	12.12	—	—	—	12.75	—	12.75
1737	12.62	12.75	—	—	—	—	—	—	12.69	12.00	—	12.12
1738	—	—	—	12.75	—	—	—	—	—	—	—	
1739	—	12.25	—	12.75	—	—	—	—	12.50	—	—	—
1740	11.75	12.75	—	—	—	—	—	—		—	—	12.25
1741	—	—	—	—	—	—	—	—	—	12.87	—	—
1742	—	13.00	—	—	—	—	—	—	—	—	13.00	—
1743	—	—	—	—	—	—	13.00	—	—	—	13.00	—
1744	—	—	—	—	—	—	—	—	—	—	—	—

1	2	3	4	5	6	7	8	9	10	11	12	13
1745	—	—	—	—	—	13.19	—	—	—	—	—	—
1746	—	—	—	—	—	13.25	—	—	—	—	—	—
1747	—	—	—	—	—	—	—	—	—	—	—	—
1748	—	—	—	—	—	—	—	—	—	—	13.50	—
1749	—	—	—	—	—	13.50	13.25	13.62	—	—	—	—
1750	—	—	—	—	—	13.50	13.25	13.50	—	—	—	—

APPENDIX B-2

Silver Value of Copper (*Takās* per rupee)

Year	*Amber* *S. Jaipur*	*Bahatri*	*Chatsu*	*Dausa*	*Lalsot*	*Malarna*	*Niwai*	*Pirayan*	*Pahari*	*Toda* *Bhim*	*Udai*	*Wazirpur*
1	2	3	4	5	6	7	8	9	10	11	12	13
1664	17.00	—	17.00	—	—	—	—	—	—	—	—	—
1665	—	15.75	16.75	16.50	—	—	16.75	—	—	—	—	—
							16.50					
1666	16.50	—	19.00	—	—	—	—	—	—	—	16.38	—
											17.38	—
1667	17.50	—	—	—	—	—	—	—	—	—	—	—
1684	—	18.50	—	—	—	—	—	—	—	—	—	—
1685	—	18.50	—	—	—	—	—	—	—	—	—	—
		19.00										
1686	—	20.00	—	—	—	—	—	—	—	—	—	—
1687	—	—	—	—	20.00	—	—	—	—	—	—	—
1688	19.00	18.00	—	—	—	—	—	—	—	—	—	—
		18.50										
1689	19.00	—	—	—	—	—	—	—	—	—	—	—
1690	18.50	—	—	—	—	18.50	—	18.50	18.50	—	—	—

1	2	3	4	5	6	7	8	9	10	11	12	13
1694	—	20.00	19.00	—	—	—	—	19.50	19.50	—	—	—
1695	—	—	—	—	—	—	—	18.50	—	—	—	—
1696	—	19.00	—	20.00	—	—	—	—	—	—	—	—
1697	—	19.50	—	—	—	—	—	—	—	—	—	—
1702	—	—	—	—	—	—	—	—	—	17.00	—	—
1703	—	—	—	—	—	—	—	—	—	17.00	—	—
1706	—	19.00	—	—	—	—	—	—	—	—	—	—
1708	—	18.00	18.50 19.00	—	—	—	—	—	—	—	—	—
1709	—	—	18.00	17.00	—	—	—	—	—	—	—	—
1710	—	17.00	17.00	—	—	—	—	—	—	—	—	—
1711	—	18.00	—	—	—	17.50 18.00	—	—	—	—	—	—
1712	18.00	—	19.00	16.50	—	18.00	—	—	—	17.75	—	—
1713	18.00	18.00	—	16.50	—	17.50 18.00	—	17.40	—	17.50	—	—
1714	—	17.50	18.00	—	18.25	18.00	—	—	—	—	—	—
1715	—	—	18.00	16.50	—	—	—	17.40	—	—	—	—
1716	18.00	17.50	18.00	—	—	17.75 17.50	18.50	—	—	—	17.50	—
1717	—	18.50	19.00	—	18.00	17.25 18.00	—	—	—	—	—	—
1718	18,50	17,50	—	16,50	19,00	18,00	—	—	—	—	—	—

1719	17.75	—	—	—	18.00	18.00	—	—	—	—	—	—
1720	17.50	17.00	—	—	17.50	17.75	—	17.00	—	17.12	17.75	17.50
1721	17.25	16.50	18.00	17.38	17.50 17.75	18.00	—	—	—	17.12	—	15.75
1722	17.50	—	—	—	17.00 17.75	17.00	18.25	—	16.75	17.32	—	16.25
1723	16.38 16.75	16.00	17.00	17.00	16.00	16.00	17.00	—	16.00	16.25	—	—
1724	16.50	17.00	16.50	—	16.50	—	—	16.87	—	—	—	—
1725	15.75	16.00	15.38	—	—	16.00	—	16.00	14.00	—	—	16.00
1726	—	15.00	16.00	—	16.00	16.00	16.00	16.50	16.00	—	—	—
1727	16.12	—	—	—	—	16.00	—	—	16.00	15.38	15.75	—
1728	15.25	15.50	15.44	—	16.25	15.50	—	—	—	—	15.50	—
1729	15.50 16.00	—	—	—	—	—	—	15.00	—	—	15.75	15.75
1730	15.00 15.50 14.75	—	16.00	15.25	16.00	16.00	—	—	—	—	—	—
1731	15.12	—	16.00	—	16.00	16.00 16.38	—	—	—	—	—	15.50
1732	—	—	—	—	—	15.00	—	—	—	16.00	15.00	—
1733	14.50 15.00	—	14.25 15.00	13.88 14.00	15.00	15.00	—	—	—	15.25	—	—
1734	16.50	—	—	15.12	15.00	14,62	—	—	14,50	—	—	—

1	2	3	4	5	6	7	8	9	10	11	12	13
1735	14.75	—	15.00	14.75	—	—	—	—	15.50	—	14.75	15.00
1736	14.25	—	14.50	—	—	16.38	—	—	—	15.00	—	16.25
1737	15.00	15.25	15.00	15.00	15.75	14.50 15.50 16.12	—	—	14.75	16.12	—	16.50
1738	15.12	—	16.00	15.00	—	15.00	—	—	—	—	—	—
1739	—	14.75	—	14.38 15.00	15.00	—	—	—	—	—	—	—
1740	17.62	15.00	16.00	—	15.00 15.25	—	—	—	—	—	—	15.50
1741	—	—	16.00	15.00	14.00	—	—	—	—	14.62	—	—
1742	—	15.00	15.50	15.00	—	—	—	—	—	—	15.00	—
1743	—	—	15.00	—	—	15.00	15.00	—	—	—	15.75	—
1744	—	—	15.75	—	15.00	15.00	—	—	—	—	—	—
1745	—	—	14.25	15.00	14.75 15.25	14.50 15.00	—	—	—	—	—	—
1746	—	—	—	15.00	15.00 14.50	15.25	—	—	—	—	—	—
1747	—	—	—	15.00	14.75 15.00	—	—	—	—	—	—	—
1748	—	—	15.00	—	15.00	14.50	—	—	—	—	14.50	—
1749	—	—	—	—	—	14.00	14.15	—	—	—	—	—
1750	—	—	—	—	—	14.00 14.88	15.38	15.00	—	—	—	—

Bibliography

The present work is based mainly on Rajasthani sources such as *arhsattas*, *muwāzanas*, *chiṭṭhīs*, *dastūr-ul 'amals* and *'amal-dastūrs*, *taqsīms*, *nirkh-bāzārs*, *jama'-bandīs*, *parwānas*, etc., and other contemporary records. These records were preserved in Rajasthan State Archives, Jaipur, when I first consulted them in the 1960's ; these are now housed in the Rajasthan State Archives, Bikaner. In the process of the shifting and re-arranging these documents, many earlier used by the present author are no longer traceable. Some categories of documents have got mixed and many have been damaged. The authorities of the Archives, however, extended full facilities to me throughout and provided micro-films of a large number of documents at my request. It is highly desirable that the State Archives, Bikaner, should devise a system of firmly numbering its documents to enable accurate reference to be made. In the absence of such a system, I have had to content myself with stating the type of document, date, and locality. In this Bibliography the major kinds of records are listed with brief notes on their nature.

Primary Sources

Arhsatta

The *arhsattas* are perhaps the most useful of the revenue records. These give abstracts of income and expenditure of each *pargana* along with details for each village. Aurangzeb in his *farmān* to Rasikdas mentions *arhsatta Hindavi* (*arhsattas* in Hindi) among the various records to be maintained in the revenue department. It is probable that the word is really from the Hindi word *arhsath*, sixty-eight, it being supposed that it was compiled from the daily entries from as many as sixty-eight (*arhsath*) documents. These papers are invariably mentioned in the *arhsattas*, though they do not actually reach the number 68. Several types of *arhsattas* were compiled, such as the *arhsatta-jama'bandi, arhsatta-mujmil, arhsatta khazāna, arhsatta bāqayāt, arhsatta-'imārat kharch, arhsatta-bhomi, arhsatta-potadār, arhsatta-ek ḥarfi, arhsatta-qarār-bāqi, arhsatta jama' kharch, arhsatta-bābat ijāra, arhsatta zamindāra bhom*, etc.

One has to be very careful in using the information provided in the *arhsattas*, particularly when one compares the revenue statistics given in them over a number of years, because the information may be incomplete at a particular paint of time and not exactly comparable with information from *arhsatta* of other years. One should moreover select an *arhsatta* for a whole year (*wājib mas bārah* or *sāl tamām*,) since sometimes the *arhsattas* are available only for a period of some months or seven months and a few days, or a single harvest.

Besides other things these *arhsattas* throw a good deal of light on such aspects as the nature of the principality of Jaipur, pattern of agricultural production and revenue-assessment; scale of tax demand ; nature of the grant under the *bhom*, *udik* and *in'ām* ; prices of food-grains and value of silver (rupee) relative to gold (*muhr*) and copper (*takā*); pattern of *ijāra*, nature of *jagīrs* assigned to the sub-assignees, *pargana* administration and commercial activities.

The following *arhsattas* have been used.

Name of pargona	*Year* (*Samvat*)
1. Amber	1720-21, 1723, 1727, 1730, 1735, 1744-47, 1749, 1751, 1752-57, 1764-67, 1769-71, 1773-77, 1779, 1781-87 and 1789.
2. Alanpur	1784
3. Akbarabad	1789
4. Ajabgarh	1783
5. Antela Bhabhra	1768
6. Bahatri (Baswa)	1722-23, 1726, 1741-43, 1745-46, 1748, 1751-55, 1759-61, 1763, 1765-68, 1773-75, 1777-99.

7. Bhadlao	1781-87 and 1789.
8. Bhusawar	1750-51, 1773- 1780, 1783-87, 1788-94, 1798-1800 and 1803.
9. Bairath	1800
10. Bahroj	1747
11. Baroda	1780.
12. Bahroar	1768
13. Bandawal	1781
14. Bayana	1750
15. Bharkol	1781
16. Bhairana	1750
17. Bhinai	1799
18. Bhadlao	1784
19. Bhagwantgarh	1720
20. Borachi (ṭappa)	1743
21. Chatsu	1721-22, 1746-47, 1750-53, 1765-72, 1776-88, 1790-95, 1797-1803 and 1805-07.
22. Chhain	1721
23. Chandwaiji	1771 and 1801
24. Dausa	1722-23, 1738, 1741, 1744-46, 1749-50, 1753-54, 1760, 1764-67, 1769, 1771-76, 1779, 1781-82, 1793-1805 and 1807.
25. Gaonri	1721-22, 1789, 1790 and 1792.
26. Geejgarh	1765, 1768-76, 1780-92 and 1800-01
27. Gudha and Salawad	1767
28. Hindaun	1729-32, 1735, 1744-45, 1768-71, 1785-90, 1794-98 and 1801-05.
29. Harsana	1780
30. Hasanpur	1770
31. Hastera	1771 and 1801
32. Ismailpur	1723
33. Jhak	1746
34. Jalalpur	1771, 1779-83, 1715-87, 1794 and 1798

35. Khohri	1771, 1782-83, 1785-86, 1790 and 1804-07.
36. Khandari	1781-82
37. Kala Dera	1804
38. Khatoo and Amarsar	1789
39. Khawa/Khuwawa	1765
40. Khilchipur	1784
41. Khilaura	1756 and 1773
42. Kathumar	1799
43. Khoh	1777, 1783 and 1801
44. Kakar	1771 and 1801
45. Lewali	1721-23, 1768-71, 1774, 1776 and 1779
46. Lahsana	1723
47. Lalsot	1743-44, 1766-83 and 1786-1807
48. Malarna	1722, 1741-42, 1747, 1752, 1756, 1765-72, 1774-80, 1782-89, 1792-94, and 1798-1807.
49. Malpura	1717, 1720-21, 1769, 1775-83, 1785-91, 1796-1805 and 1807.
50. Mauzabad/Mozamabad	1711, 1721-23, 1746, 1751-53, 1765-72, 1774, 1776, 1782-83, 1785-96, and 1798-1807.
51. Masuda	1799
52. Mandawar	1781
53. Mathuraji	1781, 1787, 1789, 1792, 1798 and 1799.
54. Mauzpur	1780
55. Mugan	1770
56. Niwai	1717-19, 1721-23, 1738, 1741, 1748, 1759, 1765-69, 1774, 1780-85, 1787-89, 1792-95 and 1801-02.
57. Nainwa	1722
58. Nadbai	1773
59. Nagar	1773
60. Naharkho	1768

61. Naraina	1779-91, 1793-94, 1800-04 and 1806-07.
62. Paota	1799, 1802 and 1804-05
63. Patan	1771 and 1801
64. Pahari	1794
65. Pinan/Pirayan	1780
66. Pragpura	1799-1806
67. Phagi/Fagui	1729, 1741, 1743, 1745-46, 1748, 1750-61, 1767-72, 1776, 1780, and 1784-1807.
68. Ramgarh	1771 and 1801
69. Rampura	1791
70. Rewari	1722
71. Sahar	1749
72. Shahpur	1771 and 1801
73. Sonkhar	1799
74. Suner	1765
75. Sawai Jaipur	1782, 1785-86, 1790-1804, 1806 and 1807.
76. Sambhar	1768 and 1770
77. Sanganer	1743, 1745-47, 1764-70, 1772-75, 1780-82, 1785-93, and 1804-5.
78. Toda Bhim	1750, 1768-83, 1789-90, 1792-94 and 1798-1800
79. Toda Rai Singh	1774-75 and 1783.
80. Toda (Toda Thek)	1747
81. Tonk	1783
82. Udai	1743, 1768-69, 1772-73, 1780-95, 1797 and 1799-1806.
83. Untgir	1792
84. Ujjain	1772
85. Umran	1770

Yāddāsht or Yādidāstī

These documents are of a very varied nature. Sometimes the *yāddāsht* was treated as a kind of *paṭṭa* granted to the *jāgīrdārs* at the time of grant of *jāgīrs* by the rulers or at the time of restoration of it to the son of a deceased *jāgīrdār*. The *yāddāsht* contains such details of *paṭṭa*, as the terms of grant, the total amount of grant and the number of horses which were to be maintained by the *jāgīrdārs*. Some *yāddāshts* have also recorded the number of ploughs and bullocks per cultivator along with their castes (see Appendix A-3). The *yāddāshtī* of *jama'wāṣil* provides detailed account of arrears (*baqāyā*) figures. Those used in this book are :

1. *Ḥaqīqatī Hal Bail Pargano* Malarna *VS*. 1721, A.H. 1072.
2. *Yāddāshtī Ḥaqīqat Hal*, *Pargana* Chal Kalyana *VS*. 1722, A.H. 1074.
3. *Yāddāshti Hal Pargana* Bawal *VS*. 1723, A.H. 1974.
4. *Yāddāshti Ḥaqiqati Hal Bail Zubāni Patel Patwārī Pargana* Chatsu *Babat VS*. 1723/A.H. 1074.
5. *Hāẓri Asāmi Wa Bail Dehai Khāliṣa Pargana* Mauzabad, *VS*. 1723/A.H. 1074.
6. *Ḥaqiqati Asāmi Wa Bail Pargana* Malarna *VS* 1723/A.H. 1074.
7. *Yāddāshti Ḥaqiqati Dehai Pargana* Pahari *VS*. 1746.
8. *Nuskho Hal Bail Pargana* Phagui *VS*. 1762/A.H. 1113.
9. *Yāddāsht i Hal Bail Pargana* Niwai *VS* 1767.
10. *Yāddāshti Naẓr Bhent Tika SRIJI Pargana Toda Bhim*, *VS*. 1802.
11. *Yāddāshti Paṭṭādār Mina VS*. 1721.
12. *Nuskho Hal Bail Pargana* Bahatri *VS*. 1762.

Taqsīm (*taksimi*)

The papers known by the name *taqsim* are referred to in the administrative and accountancy manuals of 17th and 18th centuries. The main items of information contained in these documents are the summary account of revenue in the *pargana* including its villages ; the total area of a *pargana*, area of uncultivated land (specifying separately the area under *sīr*, *kalar*, hill, nullah, etc.) and the area of cultivable land. Sometimes, the area under *punya-udik* (charitable land) is also mentioned. The total cultivable area and revenue realized (*ḥāṣil*) harvest-wise are the main items of information. The information contained in these documents is usually but not invariably for ten years (see Appendix A-6).

The following *Taqsims* and *Muwāzana dahsāla* (also *panderah salo*) documents have been used in this book.

1. *Dahsāla pargana* Antela Bhabhra *VS* 1696-1705.
2. *Pandarah salo pargano* Antelo Bhabhro *VS* 1706-1720.
3. *Dahsāla pargano* Deoti-Sanchari *VS* 1756-1765.
4. *Dahsāla pargano* Amarsar *VS* 1758-1767.
5. *Dahsāla pargano* Bahatri *VS* 1762-1771.
6. *Dahsāla pargano* Lalsot *TS* 1768-1777.
7. *Dahsāla pargano* Geejgarh *VS* 1773-82.
8. *Dahsāla tappa* Jhak, *pargano* Phagi *VS* 1773-82.
9. *Muwāzano dahsāla pragano* Naraina *VS* 1768-1777.
10. *Muwāzano dahsālo pragano* Naraina *VS* 1773-1782.
11. *Taqsim dahsāla pargano* Dausa *VS* 1787-1798.
12. *Taqsīm dahsāla pargano* Punkhar *VS* 1789-1798.
13. *Taqsīm dahiāla pargano* Geejgarh *VS* 1789-98.
14. *Muwāzano dahsāla pargano* Niwai *VS* 1789-1798.
15. *Taqsīm dahsāla pargano* Udai *VS* 1791-1800.
16. *Terij jam'a dahsāla pargano* Niwai *VS* 1794-1803.
17. *Yāddāshtī jam'a dahsāla pargano* Niwai
18. *Taqsīm pargano* Didwana A. *H.* 1042-1051.
19. *Muwāzano dahsāla pargano* Hindaun *VS* 1790-1799.
20. *Taqsīm dahsāla pargano* Udai *VS* 1815-1824.

Takhmīna

This contains estimates of revenue of villages in a *pargana* covering the details of crop-production and the revenue demand per *bīgha* in kind. The fact that the documents give the crop-yield make them especially valuable.

1. *Takhmīno pargano* Gaonri, *faṣal k͟harīf* *V*S 1722.
2. *Takhmīno bīghawār pargano* Udai, *faṣal rabi'* *VS* 1777.
3. *Takhmīno pargano* Geejgarh, *faṣal k͟harīf*, *VS* 1780.
4. *Takhmīno pargano Phagui sakh syalu VS* 1780.
5. *Takhmino pargano* Amber etc. *sakh unahlu VS* 1783.
6. *Takhmīno pargano* Malarna, *VS* 1794.

Jama'bandīs

These *jama'bandīs* give a detailed account of the amount of revenue assessed upon an individual village or *pargana* in a single harvest or both (*kharīf* and *rabi'*). The rate of assessment per *bīgha*, rate payable in cash or in kind, the total produce, the share of the cultivators and the state, showing the kind of crops and their prices, perquisites in cash or kind, the village charges and expenses, the total value of the produce in money are given in these valuable documents/(See Appendix A-2). Fortunately, a large number of documents survived for the year 1664 and 65. However, they are also available for the later period. I have used the following *jama'bandī* documents :

Pargana/Qaṣba/Mauẓa	Year (*VS*)
1. *Qaṣba* Nainwah	1717
2. *Pargana* Kotla	1720
3. Nainwa	,,
4. Mauzabad	,,
5. Malarna	,,
6. Baroda Meo	,,
7. Phagi	,,
8. Dausa	,,
9. Lihsana	,,
10. Bawal	,,
11. Nihar	,,
12. Mundawar	,,
13. Villages Akora, *pargana* Kotla	,,
14. Sakras	1721
15. Ismailpur	,,
16. Amarpuro-i *parganan* Mundawar	,,
17. Dogra, *pargana* Mauzabad	,,
18. Dwarikapur, *pargana* Kotla	,,
19. Rupila, *pargana* Udai	,,
20. Dara Shikohpura, *qaṣba* Liwali	,,
21. Ghaghoti, *pargana* Ismailpur	,,
22. Barohar	1722
23. Nainwa	,,

	Pargana/Qaṣba/Mauẓa	Year (*VS*)
24.	Udai	1722
25.	Kotputli	,,
26.	Baroda Me	,,
27.	Qasba Mauzabad	,,
28.	Lihsna	,,
29.	Baroda Rana	,,
30.	Chal Kalyana	,,
31.	Liwali	,,
32.	Malarna	,,
	Villages	
33.	Bilore, *pargana* Barohar	,,
34.	Biramwas, *pargana* Nainwa	,,
35.	Tudila, *pargana* Udai	,,
36.	Bairai, Raihans, *pargana* Kotputli	,,
37.	Muzaffarpur, Silithal, *pargana* Baroda Meo	,,
38.	*Qaṣba* Mauzabad	,,
39.	*Qaṣba* Mauzabad	1723
	Parganas	
40.	Mundawar	1746-47
41.	Jalalpur	1748
42.	Kol	1751
	Villages	
43.	Surtaniyar and Hawriya, *pargana* Phagi	1755
44.	*Pargana* Mathuraji	1789.

Dastūr-ul ʻamal and ʻAmal-dastūr

These *dastūr-ul ʻamals* (schedule of revenue rates) were framed time to time by the state. It was a body of instructions for local revenue officers. Some of the *dastūr-ul ʻamals* are preserved in bundles and mixed with other documents. References to *dastūr-ul ʻamal* have been made in *chiṭṭhīs*, *arhsattas* and *parwānas* quite frequently.

The ʻ*amal dastūrs* usually provide information or rules for the exactions and levies particularly on local trade and commerce. The *dastūr-ul ʻamal* and ʻ*amal-dastūr*, in fact, comprise together the entire set of rules and regulations providing us with a comprehensive schedules of the local revenue-rates and levies on agricultural or non-agricultural products (See Appendix A-4 and 5). The *dastūr-ul 'amals* and *'amal dastūrs* used in this book are listed below :

1. *Dastūr 'amal, pargano* Khohri — A.H. 1049
2. *Dastūr-'amal Wa 'amal dastūr Chowtra Kotwali* — 1741
3. *'Amal dastūr ḥāṣil Bhomi* — 1741
4. *Dastūr-'amal Wa 'amal-dastūr' pargano* Phagi — 1748
5. *Dastūr-'amal, ghas-charai, pargano* Chatsu — 1769
6. *Dastūr-'amal pargano* Mauzpur — 1770
7. *Yāddāshti dastūr-'amal Wa 'amal-dastūr ḥāṣil rāhdārī pargano* Phagui — 1772
8. *Dastūr-'amal pargano* Udai — 1772 ?
9. *Dastūr-ul 'amal pargano* Antela Bhabhro — 1772 ?
10. *Dastūr-'amal pargano* Bhairano — 1780
11. *Yāddāshtī 'amal-dastūr Māpa Rāhdārī, pargano* Khohri — 1781
12. *'Amal-dastūr Kotwalī Chowtra, qaṣba* Phagui — 1782
13. *Dastūr-'amal, pargano* Amarsar — 1783
14. *Dastūr-'amal Wa Mapa Rāhdārī Bhomi pargano* Hindaun — 1798
15. *Dastūr-'amal Mapa Rāhdārī, pargano* Bahatri — ?
16. *Dastūr-'amal Mapa Rāhdarī pargano* Chatsu — ?
17. *Dastūr-'amal, pargano* Jhak — ?
18. *Dastūr-'amal, ṭappa* Bawri, village Malpura — ?
19. *Dastūr-ul 'amal, pargano* Antela Bhabhro — ?
20. *Dastūr-'amal, pargano* Toda Thek — ?
21. *Dastūr-'amal, pargano* Naharkhoh — ?
22. *Dastūr-'amal, pargano* Harsana — ?
23. *Dastūr Rāhdarī. pargano* Malpura — ?
24. *Dastūr-'amal māl-o-jihāt Wa sā'ir-jihāt pargano* Didwana — ?
25. *'Amal-dastūr, pargano* Baroda. — ?

K͟hasra, K͟hasro, Kasra

The *khasra* documents specify the name of the proprietor and the cultivator of each field, its length and breadth, its area in *bīghas* and *biswas*, the quality of the soil and the crop growing upon it. The shares of the state and *ra'iyati* out of the total production is also started. The rates on different crops under *ẓabṭi* are similarly given. These papers were prepared and duly signed by the village accountant (*paṭwāri*). (See Appendix A-7). The *K͟hasras* used are :

	Mauẓa' (village)	Year (*VS*)
1.	*K͟hasro mauẓa'* Bhadlai	1718
2.	*K͟hasro mauẓa'* Sundri	1719
3.	*K͟hasro mauẓa'* Nadoti	1719
4.	*K͟hasro mauẓa'* Kawarwas *Oki Jowar mung Ki Bichoti*	1721
5.	*K͟hasro mauẓa'* Gambhira	1721
6.	*K͟hasro mauẓa'* Kalohota	1721
7.	*K͟hasro mauẓa'* Dharko Rali	1721
8.	*K͟hasro Mapa mauẓa'* Khaspura	1721
9.	*K͟hasro jama'bandi mauẓa'* Tuto *Faṣal Rabī' pargana* Tonk	1768
10.	*K͟hasro mauẓa'* Baron *Faṣal Rabi'*	1769
11.	*K͟hasro mauẓa'* Ghanghka *Faṣal Rabī'*	1769
12.	*K͟hasro manẓa'* Chimanpura Vas Deoli	1785
13.	*Faṣal Kharīf, pargana* Lalsot	
13.	*K͟hasro mauẓa'* Ratanpura, *Faṣal Kharif*	1785
14.	*K͟hasro mauẓa'* Bilka, *Kharif, pargana Lalsot*	1785
15.	*K͟hasro mauza'* Naeela, *Sakh* Unahlu	1791
16.	*K͟hasro mauẓa'* Mohanpura, *ṭappa* Khoh *ẓabṭi, sakh syalu*	1791
17.	*K͟hasro mauẓa'* Rohtaspur, *sakh syalu*	1793
18.	*K͟hasro ẓabṭi mauẓa'* Mohanpur, *sakh syalu*	1795
19.	*K͟hasro mauẓa'* Sigarpura Vas Asarpur, *Sakh Unahlu*	1100
20.	*K͟hasro mauẓa'* Huchukara, *sakh syalu*	1802
21.	*K͟hasro mauẓa'* Bokarabas	1802

22. *Khasro mauza'* Gazi Ka Thana, *sakh syalu* 1802
23. *Khasro mauza'* Morara, *pargana* Naraina 1802
24. *Khasro mauza'* Bhangarh, *sakh unahlu* 1817
25. *Tazkaro mapāi, mauza'* Kakso, *pargana* Baraud 1836
26. *Khasro, jama'bandi, mauza'* Madsudanpura, *pargana* Chatsu 1837
27. *Khasro qaṣba* Syobdaspur, *pargana* S. Jaipura 1848
28. *Khasra jama'bandi, Faṣal Kharīf, pargana* Chatsu 1853
29. *Khasro mauza'* Hirapuro, *sakh unahlu, ṭappa* Khoh, S. Jaipur 1855
30. *Khasro mauza'* Akheypur Vas Akora, *sakh syalu, ṭappa ḥaveli, pargana* Sawai Jaipur 1857
31. *Khasro qaṣba* Vakra 1864
32. *Tazkaro mapāi mauza'* Piplod. 1865

Nirkh Bāzār

These records give the market rates prevailing in a *pargana* or township (*qaṣba*) for all kinds of commodities. It included the prices of agricultural and non-agricultural products along with gold (*kundan*), gold *muhr*, silver and different kinds of *takā* (copper). The list of *nirkhbāzārs* used is given below.

Qaṣba Pargana :	*Year (VS)*
1. Mauzabad	1721-22, 1777, 1791 and 1792
2. Chatsu	1722-23, 1769, 1776, 1781-82, 1785, 1788 and 1807.
3. Sanganer	1721, 1774, 1776-77, 1780, 1782, 1784 and 87.
4. Liwali	1722
5. Khohri	1722, 1782 and 1785.
6. Punkhar	1722, 1762, 1771, 1779-80, 1783-84 and 1795.
7. Khilohra	1723
8. Niwai	1723, 1773, 1779-80, 1783, 1800 and 1806-07.

9. Udai	1723, 1750, 1773, 1777, 1780, 1782, 1784-85, 1789, 1792, 1799-1800, 1805.
10. Bahatri	1747, 1771, 1773, 1778, 1783, 1785, 1796-97 and 1799.
11. Kot	1747
12. Pahari	1747, 1779, 1782, 1791-92, 1794 and 1796.
13. Sherpur	1750
14. Bhusawar	1750 and 1773-74.
15. Pirayan	1752, 1769-70, 1772, 1777, 1781-83, 1786 and 1806-07.
16. Toda Bhim	1759-60, 1769-70, 1777-80, 1784, 1789-90, 1793-94 and 1798.
17. Amarsar	1759, 1769-70 and 1779-80.
18. Pragpura	1762, 1785, 1790, 1794, 1797, 1801-02 and 1805-07.
19. Salawad	1769 and 1779.
20. Amber	1770, 1776-78, 1780-82, 1785-88, 1791 and 1797
21. Phagi	1770, 1778-79, 1783, 1785, 1787, 1789, 1805 and 1807.
22. Malarna	1770, 1780, 1782, 1788, 1791, 1793-95, 1802-03 and 1806-08.
23. Lalsot	1771, 1778-80, 1785, 1797 and 1801-04.
24. Lalgarh	1772
25. Suner	1774
26. Tonk	1774, 1783, 1788, 1796-97.
27. Gazi Ka Thana	1777, 1785 and 1787.
28. Hindaun	1777, 1786, 1788, 1800-01 1804 and 1807.
29. Paota	1778, 1790-91 and 1807
30. Ajabgarh	1777, 1780-81, 1783-85 and 1790
31. Bhangarh	1777-79
32. Dausa	1778, 1780, 1790-92 and 1796-97
33. Wazirpur	1778-80, 1782, 1786, 1788, 1792-94 and 1797
34. Geejgarh	1779, 1781, 1788-89, 1792-93 and 1797-98

35.	Chandwaiji	1779-82.
36.	Jaitpur	1780-82
37	Khatoo	1780, 1783 and 1793
38.	Bairath	1781
39.	Sanganer	1782 and 1784
40.	Rini	1782, 1795-97 and 1799
41.	Manoharpur	1786, 1796 and 1798
42.	Naraina	1787, 1804 and 1807
43.	Punkhar	1788 and 1796
44.	Patan	1792
45.	Jalalpur	1796
46.	Rampura	1797-99 and 1805.

Nirakh-Jamaʿbandī

These rather rare documents are supplementaries to *nirkh-bāzār*. They give current prices of different crops datewise under *kharīf* and *rabīʿ*. The records are maintained for each village. I have consulted *nirkh-jamaʿbandī* of *pargana* Bahatri for the year, *VS*. 1723/1664.

Sanads

A large number of documents are maintained under this head. They contain miscellaneous information such as an individual's emoluments, titles, privileges, offices or the state rights to revenue from land. The *sanads* were issued by the *dīwān* and a record of them was kept in his office. These are, of course, too numerous to be listed here.

Parwāna

There is a separate category of these documents preserved in the Archives. The *parwāna* was an order or letter of instructions issued by the Mahārāja generally to his *dīwān*, *vakil* and other subordinates.

The *parwānas* acquaint one with the rights and duties of *diwān* and other *pargana* officials. Thc *parwānās* (*ṣabti*) provide us important information about lands or villages granted in *punyaudik* to the Brahmans taking care of the temples or to the *charans* or *bhats*. Sometimes these grants were made on copper plates (*tambra-patra*) and a *sanad* of it was maintained in the office of the *dīwān*. As with *sanads*, so with *parwānas*, owing to the large number that have been used, no list can be furnished here.

Muwāzana

These documents contain the history of various villages in a *pargana* for a very long period of years. They are arranged in strict alphabetical order and were compiled in the 19th century with the help of other documents like *raqba-bandi* and *athsattas* of different kinds. They show the number of villages, the area of arable land, land under cultivation, waste, forest, pools, nullah, hilly strips. These also provide us with the details of *jama*ʻ figures. Since we do not have a continuous series of *arhsattas*, these documents sometimes fill the gaps. The detailed information as to whether a particular village, in a particular year was in *khaliṣa* or *jāgir* or *in*ʻ*ām* or *ijāra*, is contained in such documents.

These documents often contain certain lacunae. For example, the figures of *ḥal-ḥaṣil* are collected from *arhsattas* of different categories and set side by side, though for reasons stated above, they are not exactly comparable.

Muwāzana Kalān and Muwāzana K͟hurd (k͟hiṭṭhīs)

These records are a collection of loose papers (many thousands in number) kept in bundles on the basis of years. These small-size papers referred to as *chiṭṭhīs* are valuable for understanding the working of the institution of *zamīndāri*, containing as they do valuable information about the condition of peasants, forced labour (*begār*), complaints of peasants lodged with the *pargana* officials about the oppression of *zamindārs*, and *jāgirdārs*. An assessment of the state policy towards the *zorṭalabi zamindārs* and *jāgirdārs* can be made from such *chiṭṭhis*. The land rights of different classes and the obligation of landed class towards the state are recorded in miscellaneous papers contained in these bundles.

Awārija or Awārijo

Literally *awārija* is a diary, a day-book, a ledger, a rough note book, an abstract, an account of receipts and disbursements. The *awārijas* are of different types such as *awārija-nota*, *awārija-peshkash paṭwārī*, *awārija-distūr-chaudhurī*, *awārija-bāgi awārija-khāṣ*, *awārija-tāqai*, *awārija-mutāliba*, *awārija-muqta'i*. *awārija-qarẓ*, *awārija-wajah yāft* etc. I copied out the full text of *awarija mutaliba* of *pargana* Bahatri *VS*. 1749/1694 and found it to be very important. It shows the amount of arrears outstanding against the *jāgirdārs* (Mahārāja's sub-assignees) towards the rulers, the number of contingents to be maintained by them and their salaries.

Roznāma, Roznāmcha

These are the daily account of receipts and expenditure, land revenue of each village in a *pargana* along with the different cesses other than land revenue such as *peshkash*, *rāhdāri puro*, *jurayan*, *chobi*, etc. These records were maintained by the *potadār* of a *pargana*. Unfortunately, these documents are not serially arranged.

Other Documents

1. Records from *Diwān-i-Huzūri* such as la(II), 7-E, 14-A, 15-A, 34, 39-A, 47, 74.
   Original *Lakhtangs* : *VS*. 1747, 1778, 1780, 1787, 1797, 1800 and 1802.
   Copy of *Lakhtangs* : *VS*. 1768, 1792, 1794, 1797, 1805
2. *Old Historical Records* : (Persian and Hindi ; dated and undated).

   (i) Miscellaneous Persian Papers concerning the Mahārāja of Jaipur.

(ii) Different '*Arzdāshts* written by Pancholi Jagjiwan Das, Pandit Megh Raj and Bhikhari Das.

3. *Vakil's Report* (Hindi and Persian *VS*. 1707-1715), especially Register nos. 22, 73, 75, 139, 211, 396 and 425.
4. *Fards*
5. *Siyaha Ḥaqiqati pargana* Geejgarh *VS*. 1781 and 1787.
6. *Baqāya VS*. 1758 onwards.
7. *Fehrist Deh-ba-Dehī* of different *parganas*, *VS*. 1722.
8. *Iltimāsh*-Dated and undated, Register No. 124.
9. *Khutūt-i-Mahārājgān VS*. 1701-1807.
10. *Muchalka* (Hindi) *VS*. 1767.
11. *Muchalka* (Persian) *VS*. 1805.
12. *Khuṭūṭ-i Ahalkārān VS*. 1690-1807.
13. *Kharitas* (*VS*. 1719 onwards).
14. *Yaddāshti Nazar Bhent Tika, Pargana* Toda Bhim *VS* 1802.
15. *Raqbo Pargano* Mauzabad *Sāl Samvat* 1698 (See Appendix 1-A).

B. Contemporary Works

16. Abū-l Fazl, *Ā'in-i Akbari*, Vol. II, ed. H. Blochmann, Bib. Ind. Calcutta, tr. Jarrett, revised by J. Sarkar, Calcutta, 1949.
17. Abū-l Fazl, *Akbarnāma*, Vol. II, tr. H. Beveridge, Vol. II, Bib. Ind., Calcutta, 1897-1921.
18. Jahāngīr, *Jahāngir-nāma* or *Tuzuk-i Jahāngīri*, tr. H. Beveridge, 3rd edition, New Delhi, 1978.
19. Shahnawāz Khān, *Ma'-asir-ul Umarā*, tr. H. Beveridge, *et al*., Vol. II, R.A.S., Calcutta.
20. Muḥammad Hāshim Khāfī Khān, *Muntakhab-al Lubāb*, Vol. II, ed. K.D. Ahmad and Haig, Bib. Ind., Calcutta, 1860-74.
21. Bhīmsen, *Nuskha-i Dilkusha*, MS Br. Mus. Or 23, rotograph in the Library of Department of History, AMU.
22. Munshi Nand Rām Kāyastha Srivāstavya, *Sīyaqnāma*, litograph, Nawal Kishore, Lucknow, 1879.

23. *Waqai-Sarkār Ajmer wa Ranthambore*, MS Asfiya Library, Hyderabad, transcript in the Department of History, Aligarh, Nos. 15 & 16.
24. *Catalogue of Tod's Collection*, Br. Mus., London.
25. Pelsaert, Franciso, *Jahangīr's India*, tr. Geyl and Moreland, Cambridge, 1925.
26. Muhnot Nainsī, *Mārwār-rā-parganā rī Vigat*, ed. Narain Singh Bhati, Vols. I-III, Jodhpur, 1968-69, 1974.
27. Bernier, F., *Travels in India*, tr. Constable, rev. by V.A. Smith, London, 1925.

C. Modern Works

Books:

1. *A Descriptive List of the Farmans, Manshurs and Nishan*, R.S.A., Bikaner, 1962.
2. *A Descriptive List of Vakils' Report*, 2 vols, R.S.A., Bikaner, 1967.
3. *A Descriptive List of Arzdasht* (*Rajasthani*), R.S.A., Bikaner.
4. Ali, Athar, *The Mughal Nobility under Aurangzeb*, Bombay, 1966.
5. Bhatnagar, V.S., *Sawai Jai Singh*, Jaipur, 1972.
6. Bloch, Marc, *Feudal Society*, London, 1967.
7. Chandra, Satish, *Parties and Politics at the Mughal Court*, Aligarh, 1959, 2nd ed., New Delhi, 1963.
8. ———, *Medieval India : Soeiety, The Jagirdari Crisis and the Village*, New Delhi, 1982.
9. Chicherov, A.I., *Indian Economic Development in the 16th-18th Centuries—Outline History of Craft and Trade*, Moscow, 1971.
10. Habib, Irfan, *The Agrarian System of Mughal India (1556-1707)*, Bombay, 1963.
11. ———, *The Atlas of The Mughal Empire*, New Delhi, 1982.
12. Hasan, Ibn, *Central Structure of the Mughal Empire*, London, 1936.
13. Hasan, S. Nurul, *Thoughts on Agrarian Relations in Mughal India*, New Delhi, 1971.

14. Irvine, W., *Later Mughals*, Calcutta, 1922.
15. Malcolm, Sir John, *A Memoir of Central India*, 2 vols., London, 1824.
16. Moreland, W.H., *The Agrarian System of Muslim India*, Cambridge, 1920, reprinted Allahabad.
17. ———, *India at the Death of Akbar*, London, 1920.
18. ———, *From Akbar to Aurangzeb*, London, 1923.
19. Powell, B.H. Baden, *The Land System of British India*, 3 Vols. Oxford, 1982.
20. Ranawat, Manohar Singh, *Itihaskar Muhnot Nainsi aur Usaka Itihas* (Hindi), Jodhpur, 1981.
21. Saran, P., *The Provincial Government of the Mughals (1526-1658)*, Allahabad, 1941.
22. Sarkar, Jadunath, *Mughal Administration*, Calcutta, 1920.
23. ———, rev. and ed. Raghubir Singh, *A History of Jaipur (c. 1503-1938)*, 1984.
24. Sarkar, Jagdish Narain, *Studies in Economic Life in Mughal India*, Delhi, 1975.
25. *Selected Documents of Aurangzeb's Reign, 1659-1706*, ed. Yusuf Husain Khan, Hyderabad, 1963.
26. Sharma, G.D., *Rajput Polity*, New Delhi, 1977.
27. Sharma, M.L., *History of the Jaipur State*.
28. Sharma, S.R., *Mughal Government and Administration*, Bombay, 1951.
29. Siddiqi, Noman Ahmed, *Land Revenue Administration under the Mughals (1700-1750)*, Bombay, 1970.
30. *The Cambridge Economic History of India*, ed. Tapan Ray Chaudhuri & Irfan Habib, Vol. I, c. 1200-1750, 1982.
31. Watt, G., *The Dictionary of the Economic Products of India*, 6 vols, London, 1889-93.
32. Wills, C.U., *The Land Tenures and Special Power of Certain Thikanadars of Jaipur State*, 1933.
33. Wilson, H.H., *A Glossary of Judicial and Revenues Terms*. London, 1875.

Articles

1. Bhadani, B.L., 'The Allodial Proprietors ? : The Bhumias of Marwar', *IHR*, Vol. VI, No. 1-2, July 1979-Jan. 1980.

2. Chandra, Satish & Gupta, S.P., 'Jaipur Pargana Records', *IESHR*, Vol. VI, No. 3, Sept. 1966.
3. ———, & Singh, Dilbagh, 'Structure and Stratification in the Village Society in Eastern Rajasthan', *PIHC*, Muzaffarpur, 1972.
4. ———, 'Some Aspects of Indian Village Society in Northern India during the 18th Century', *IHR*, I, 1974.
5. ———, 'Some Aspects of the Growth of a Money Economy in India during the 17th Century', *IESHR*, III (4), 1966.
6. ———, 'Some Institutional Factors in Providing Capital Inputs for the Improvement and Extension of Cultivation in Medieval India', *IHR*. Vol. III, No. 1, 1976.
7. Grover, B.R. 'Nature of Land Rights in Mughal India', *IESHR*, I, 1963.
8. ———, 'Nature of the Dehat-i Taaluqa (*zamindari* villages) and the evolution of the taaluqdari system during the Mughal Age', *IESHR*, II, 1965.
9. Gupta, S.P., and Shireen Moosvi, 'Bhomi in the Territories of Amber (c. 1650-1750)', *PIHC*, Jabalpur, 1970.
10. ———, & Shireen Moosvi, 'Weighted Price and Revenue-Rate Indices of Eastern Rajasthan (c. 1665-1750)', *IESHR*, XII, 1965.
11. ——— & Dilbagh Singh, 'Famines in the Territories of Amber (c. 1650-1800)', *PRHC*, Ajmer, 1975.
12. Gupta, S.P., 'Expansion of the Kachhwaha Territory in Mughal Times', *PIHC*, 1965.
13. ———, '*Ijara* System in Eastern Rajasthan (c. 1650-1750)', *Medieval India—A Miscellany*, Vol. II, 1972.
14. ———, 'The System of Rural Taxation in Eastern Rajasthan (1965-1750)', *PIHC*, Muzaffarpur, 1972.
15. ———, 'Jhain of the Delhi Sultanate', *Medieval India—A Miscellany*, Vol. III, 1973.
16. ———, 'New Evidence on Agrarian and Rural Taxation in Eastern Rajasthan (17th & 18th c.)', *PIHC*, Aligarh, 1975.
17. ———, 'A Note on *Khasra*', *Medieval India—A Miscellany* Vol. IV, 1976. Reprinted from *Madhya Kalin Bharat*, Part II, 1983, ed. Irfan Habib.
18. ———, 'Agrarian Information in *Taqsim* Documents from

Eastern Rajasthan (1649-1767), *PRHC*, reprinted from *Madhya Kalin Bharat*, Part I, ed. Irfan Habib, 1981.

19. ———, 'The Village and Further Evidence of the Stratification of Village, Peasantry in Eastern Rajasthan (c. 1665-1808)', *Jagdish Singh Gahlot Commemoration Volume* on Some Aspects of Socio-Economic History of Rajasthan ed, G.L. Derva, 1980.
20. ———, 'Agrarian Condition in Mid-17th Century Mughal India—An Official Survey of Pargana Mauzabad, 1641, *PIHC*, Bodh-Gaya, 1981.
21. ———, 'A Note on *Palti* from Jaipur Region', *PRHC*, 1982.
22. ———, 'Prices and Rural Commerce in 17th Centuries Eastern Rajasthan', *PIHC*, Durdwan, 1983.
23. ——, 'Prices in the Trans-Yamuna Tract from the Close of the 17th to the Later Half of the 18th Century', *PIHC*, 1984.
24. Habib, Irfan, 'Potentialities of Capitalistic Development in the Economy of Mughal India', *Enquiry*, Vol. III, No. 3, 1971.
25. ———, 'Usury in Medieval India', *Comparative Studies in Society and History*, Vol. VI, 1964.
26. ———, 'A System of Trimetallism during the Price Revolution'—A Study of the Extent of the Use of Silver-Money in the Mughal Indian Economy, paper presented at a Conference : The Monetary System of Mughal India, held at National Humanities Centre Research Triangle Park, North Carolina, U.S., 3-5 June 1981.
27. ———, 'The Currency System of the Mughal Empire (1556-1707)', *Medieval India Quarterly*, Vol. IV, Nos. 1-2, Aligarh, 1960.
28. ———, 'The Peasant in Indian History', *PIHC*, Kurukshetra, 1982.
29. ———, 'Distribution of Landed Property in Pre-British India', *Enquiry*, N.S. II (3) (O.S. No. 12), Winter, 1965, Delhi.
30. Halim, Sumbul, 'Vakil Reports in Rajasthan State Archives—A Sample Study', *PIHC*, Annamalainagar, 1984.
31. Hasan, S. Nurul, '*Zamindars* under the Mughals', *Land*

Control and Social Structure in Indian History, ed. L.E. Frykenberg, London, 1969.

32. ———, K.N. Hasan & S.P. Gupta, 'The Pattern of Agricultural Production in the Territories of Amber', *PIHC*, Mysore, 1966.
33. ———, & S.P. Gupta, 'Prices of Food Grains in the Territories of Amber', *PIHC*, Patiala, 1967, Part I, Patna, 1968.
34. Husain, Afzal, 'Marriages Among Mughal Nobles as an Index of Status and Aristocratic Integration', *PIHC*, Muzaffarpur, 1972.
35. Khan, Iqtidar Alam, 'The Middle Classes in the Mughal Empire', *PIHC*, Aligarh, 1975.
36. Moreland, W.H., 'The Development of the Land Revenue System of the Mogul Empire, *JRAS*, 1922.
37. Mukhia, Harbans, 'Illegal Extortions from Peasants, Artisans, and Menials in 18th Century Rajasthan', *IESHR*, Vol, XIV, No. 2, April-June 1977, New Delhi.
38. Rana, R.P., 'Agrarian Revolts in Northern India during the late 17th & early 18th Century', *IESHR*, XVIII (3 & 6), 1981.
39. Sharma, G.D., 'Indigenous Banking and the State in Eastern Rajasthan during the 17th Century', *PIHC*, Waltair, 1979.
40. Siddiqi, N.A., 'The Classification of Villages under the Mughals', *IESHR*, I (3).
41. Singh, Dilbagh, 'Caste and Structure of Village Society in Eastern Rajasthan during the Eighteenth Century', *IHR*, Vol. II, No. 2, Jan., 1975.
42. ———, 'Nature and Incidence of Taxes levied on Inland Trade in Eastern Rajasthan during the 17th & 18th Centuries', *PIHC*, Bhubaneshwara, 1977.
43. ———, 'The Role of Mahajans in the Rural Economy in Eastern Rajasthan during the 18th Century', *Social Scientists*, No. 22, May 1974.
44. ———, 'Position of Patel in Eastern Rajasthan during the 18th Century', *PIHC*, 1970.
45. *Vigat* No. 9 : Project, 'Preservation, Listing Cataloguing, Editing, and Publication of Manuscripts', University of Rajasthan, Jaipur, (cyclostyled).

Additions and Corrections

Note: Lines are counted, unless otherwise indicated, from beginning and end of main text only.

P. 39: L. 14. For '*paragana*' read '*pargana*'.
P. 59: L. 7 from bottom: For 'Bhatri' read 'Bahatri'.
P. 62: L. 13 from bottom for '*Chombi*' read '*Chombli*'.
P. 68: L. 5. For '*Aheer*, read 'Ahīr'.
P. 89: L. 19. For 'ending 1717' read 'ending in 1717.'
P. 104: Add at the end of f.n. 11: 'Tapan Raychaudhuri and Irfan Habib, '*The Cambridge Economic History of India, Vol. I, c. 1200-1750*,' pp. 166-71, 1982.
P. 104: L. 18 and f.n. 11: L. 1. For '(290-296)' read '(310-16)'.
P. 107: f.n. 17, should be deleted.
P. 108: f.n. 18: For no. 18 read 17. Also in the text for f.n. 18 read 17.
P. 112: L. 4 from bottom: For f.n. no. 1 read 18.
P. 112: Add at the bottom, the following as footnote 18: '*Nakhkhas* was a market for horses and cattle'.
P. 114: L. 12 from bottom: Read the last word as '*pargana*.'
P. 119: L. 6 from bottom: Read *pa'is*.
P. 135: L. 17: For 'Shaikhwats' read 'Shaikhawats'.
P. 135: L.8 from bottom: For '*jamīndārī*' read '*zamīndārī*'.
P. 136: L. 4 from bottom: For '*paragana*' read '*pargana*'.
P 137: f.n. 16: L. 2 & 4 For: Narula' read 'Naruka'.
P. 142: L. 8: For '*bohras*' read '*bohra*'.
P. 142: f.n. 35: For '1723/1766' read '1723/1666'.
P. 146: L. 11: For 'quallty' read 'quality'.
P. 151: L. 21: Read '*sarhi*'.
P. 151. L. 19: For '*baqāyat*' read '*bagāyat*'.
P. 152: Last line of the text: Transfer f.no. 27 from *Khandsāl* to *agarkhāri*.

P. 153: L. 2: For 'Karwan' read 'Caravan'.
P. 165: f.n. 12: l. 5. For '*Chautho*' 'read '*chautho*'.
P. 167: L. 6: For 'Bishen' read 'Bishan'.
P. 170: L. 14: For '*gunjaish*') margin' read '*gunjaish* '(margin
P. 172: L. 9: For 'Sawat' read 'Sawai.'
P. 175: L. 4: For '*chaudharies*' read '*chaudhurai*'.
P. 185: f.n. 118: L. 2 from bottom. For '*peshaksh*' read '*peshkash*'.
P. 190: L. 2: For (Abachal Ram read (Abachal Ram).
P. 225: L. 17: For *jāgīrdā* read *jāgīrdār*.
P. 240: Title of the book at top. For 'Easrern' read 'Eastern'.
P. 255: L. 5 from bottom. For 'no' read 'on'.
P. 296: Top title of the book. For 'Ararian' read 'Agrarian'.
P. 296: L. 2 (Hindi). For *kola Tori Riwani* read *Ko Lato dori vani*.
P. 318: L. 10: For 'paint' read 'point'.
P. 318: L. 17 from bottom. For '*jāgris*' read *jāgīr*.
P. 323: L. 13: For dahiala' read '*dahsāla*'.
P. 327: L. 5: For 'started' read 'stated.'
P. 331: L. 8 from bottom. For '(*khittis*)' read '(*chiṭṭhīs*)'.

Index

Abu'l Faẓl, 2
agarkhāri, 152 & n
Agra (Akbarabad), suba, nos. of *pargana* & vills. in *J*, 28-30; trade in, 110-12 & 115, 121, 289, 318
Agrarian society, 116-33
Agriculture, 38, 122
Agricultural produce, 40 & n; revenue derived from, 45-50; proportion of revenue, 64-66; produce per *bīgha* and total, 69-73; trade in, 93 & 101-04 & 108-14; consumption of, 93 & 100-03, 141
Ahīr, 67, 131; caste, 181
A'itiqād Khān, 15
Akbarnagar, 6
'alūfa, *'alūfo*, 172, 174
'amal, 139
'Amal dastūrs, 106-07, *chabūtra kotwālī*, 108-10, 269, documents of, 270-79
Amar Singh, 221-22
Amarsar, *pargana*, 15, 43, 124, 145, 166n; *dastūr-ul-'amal*, 326; *nirkh bazār*, 329
Amber, *pargana*, 1-5; expansion and rulers of, 5-27 & 36; vills. held in *J.*, 28-37; territories held by the rulers of, 37, 38; revenue demand in, 59, 94; prices of food-grains of, 83, 84; indigo, 86; *jinsī* commuted into cash of 100; 135, 137, 140; *sāi'r-jihāt* of 150-51; proportion of *muwāfiq* and *siwāi jama' bandī* of 160, absolute figures, 160, 163; *dīwāns* 186; cf. *'āmils*, *amīns* and *kārkūns*, 188-90; villages under *khāliṣa* and *jāgīr* of, 213; Terij Shekhawat, 221, 224, 229, 232; silver value of gold of 313-16; *Arhsattas*, 318; *Nirkh bāzār*, 32. See Sawai Jaipur.
'āmil, revenue collector and assessor, 164 & n—74 & n; and *amīn*, 172; appointment and obligations of, 166-74; pay of, 173, 178; names and tenures of, 188-210. See *amīn*
'āmilāna, 14n, 138n
amīn, revenue assessor, 164 & n—74 & n; appointment and tenure of, 188-210
Anglo-Jaipur Treaty of, 1811, 230
Antela Bhabhra, *pargana*, 24; area of, 43, 120; *taqsīm pandrahsāla* of, 289; *Arhsatta* of, 318; *Dahsāla* of, 323; *Dastūr-ul-'amal* of, 326. *ārāẓī* of, 41, 233
Arhsatta, pattern of cropping, 38-9; 39; area under cultivation, 41-73; prices in, 74 & 83 & 85-6 & 317-21
āsāmī, 63; castes of 67-8, 118, 120; taxation upon and bullocks of, 126-33, 184; horses held by 221-22, 294; area under cultivation of, 296-303, 322
aṣlī, vills., 116
Aurangzeb, 6, 12, 75, 86, 224
Au, *pargana*, 8, 102
awār, *awārdar*, 93
awārija, *awārijo*, 165, 219n; nature of, 322

'Āẓam Khān, Shāh, 2, 12, 15
'Azīzuddīn, Bādshāhzāda, 16
bāgdam, 109
bagoti, *bagāyat*, 141, 151-52

Bahadur Shah, 163
Bahatri (Baswa), *pargana*, 7 & n, 8, vills., in *J.*, of 11n; change in the number of vills., 39; revenue from, 46-7 & 49-50, 52; revenue demand in, 57-8; state share of produce, 59, 95; percentage of area under different crops, 61-2; proportion of revenue from *kharif* and *rabi'*, 65; area, production and prices of, 69-73; prices, 83; proportion of *bichotī* and *anya*, 100, 121, 135; *sai'r-jihāt* of, 150-01; *muwāfiq-jama-'bandi* and *siwāi-jama'bandī*, 160, 173 & n, vills., in *khāliṣa* and *jāgīr*, 213, 229
baijhri, 62
baithak, cess, 106
baksi, Pers. *bakhsi*, 169n, 214
baitulmāl, 153 & n
bājra, millet, 40n; revenue from, 45, 48; State share in, 59; area under, 61, prices of, 79 & 83
Banawar, *pargana*; revenue statistics, 40n
bandh-birar, 182 & n
bandhati, cess, 126
bandūkchī, *bandūqchī*, 214
balad, 165
bania(s), grain merchant, 93 & 100-02 ; money-lender, 105, State share from, 126
banjār, land, 43, 59
banjārās, nature of, 105; grain merchants, 109 & 111-12, 168
bāqī, 166-67, 177, 244, 249, 254, 256
baqāyā, 170 & n. See *baqāyā*,
bāra, *vāra*, kyari, 148 & n
Barguzar, caste, 139n
Barley, crop; 40n; revenue derived from 48-50; State share in, 59; area under 61-2; prices of 78 & 83 ; sale of, 101-02
barti, millet, 60
Basti, 41
Baswa, *pargana*. See Bahatri
baṭ'ā'ī a system of assessment, 40, 44, 128-29, 145 & 179; absolute figures of, 245-49. See *jinsī*
batwāli, cess, 106, 111
Bawal, *parganai*, 6
Bayana, *pargana*, 22, *J.* of 31, 115, 319
baithi begār, 123n
bāz-yāft, *wajah yāft*, 332
Begār, 122, 123n, 331. See *jhupri*
Bengal, 6
Bernier Francois, 155
Bhagwandas, 6, 38n
Bhairana, *pargana*, 15, 126, 319, 326
bhaoli, *bhaoli baṭ'āī*, 280, 282, 286. See *batā'ī jinsī* & crop-sharing
Bhao, Singh, 3
Bharkol, *pargana*, 6, 121, 319
Bharamal, Raja 1-2, 5-6, 175
bhent, 151, 153 & n
bharoti, 107
Bhartha, 114
Bhayya, kasti (*kayath* or *kayastha*), 174
Bhikari Das, 114
bhom (*bhūm*), 134n, 135; a cess, 137-38
bhom-bali, a cess, 137n
bhomia, 123, 127, 134 & n; *zamīndārs* castes of 135; Rights and obligations, 136-40, 137n, 154, 168, oppression of 137n-139 & n
bhomichara, 134n
bichhaiti, *bachhayat*, meaning of, 105; Incidence of taxation, 108-13
bichotī, 40, 93, 100-02, 269
Bidar Bakht, 12

bīgha, 40 & n, 41, 43-4, 50, 60, 69-71, 290, 323-24, 327
bīgha,-i daftarī, 41-43
bīgha-i Ilāhī, 41, 43
Bishan Singh, Maharaja, *parganas* held in *jāgīr* of, 8, 11n, 120; 139, 167, 175
bisondh, 183
biyaz-ghiwai, 142
bohrās, status and obligations, 139-43; money-lender, 141-42; privileges, 142, 167; guarantor, 228n
Brahmans, concessional revenue-rates of, 123-25; peasants, 130; bullocks/cattle held by, 131-33, privilege class of 146 & 149; *pargana* officials, 174; as *paṭels*, 181
Bullock, 63, rate per, 113; number, 67-8; 131-33, 131n, 257-68
Bullock carts, 113
Butter. See *ghī*
bıopāri, vyopāri, merchant, 105, kinds of merchants, 105, 107

Camels, 112
Caravans, *qāfilas*, 110-12.
Cash crops, 62; area under cultivation, 61-2; rate of revenue demand of, 55-9
chiṭṭhī, 109, def. & rate of, 156 & 159
chakbandi, 171 & n, 182
Champavat, Fateh Singh, 232 & n
Chandra, Satish, 12, 32n, 63, 119, 120n, 122n
Chatsu, *pargana*, vills., held by Maharaja, 11 & n, 12, 4; assignees of, 18, vills., held in *jāgīr*, 34, 39; measured area, 42; revenue from, 45-6 & 49-50, 53, 57-8; State share, 59, 60, 96; area under different crops, 61-2, 63, revenue from *kharīf* and *rabī'*, 66; prices of food-grains, 83-85; agrarian trade in, 113; *āsāmīs* and bullocks of, 120 & 133, 140; *sair-jihāt* of, 150-51, 154, 157; *Muwāfiq jama'bandī* and *Siwāi'-jama'bandī* of, 160-61; pay of *pargana* officials of, 173; names of *pargana* officials, 202-05; 212; vills., under *khāliṣa* and *jāgīr* of, 213, 229; ploughs and bullocks, 255 & 257-68; measured area, 256; silver value of gold and coppe:, 310-16, 319; *Dastūr-ul-'amal*, 326; *khasra* 327; *Nirkh bāzār*, 328
Chaudhurī, 121; rate of taxation, 124-26, 127, 148. 164; appointment of, 167, 168, 170; Duties and privileges of, 174-76, 182, 228. See *qānūngo*
Chaudhuri, Tapan Ray, 104n
Chaukīs, 103, 110-12
Chauhans, 135
Cheena, *arzan*, 62
Chhan, Chhain, *pargana*, revenue of, 22, vill. of, 34 & Remarks, 34, 319
Chimaji, 17
Chipro, 131
chiṭṭhīs, concerning Land Revenue Administration, 163-86 & n
Chobdar, 103
Chombli, 62, revenue demand, 49-50
Chowkayat khazāna, 145
Chowkīdārī, 152
Commodities, 108-13
Copper; Silver price of 89 & 313-16; Indices of prices of, 91
Copper money, 108n
Cotton, *vani*; area under, 40n & 61-2; revenue from, 45-8, 55-9; 62; cultivation of, 60; trade in, 109-11
Cows, 159
Credit, 141-42
Cropping, 42-3 & 43-50
Crop-rate, 55-9, 69-73, 84

Crop-sharing; State share in, 59, 64-6, 124-26, 128-29; see *baṭā'ī*
Cultivation, extent of, 41-3; 44-50

dah-nīmī, def. & rate of, 145 & n
dah-sāla papers, 41n, 116n, 289-93, 323; see *taqsīm* and *muwazana-i dah-sāla*
dākhili, 116
dalāl, 106, 110
dām, copper money, unit of account, 74n, 108 & n, 108-13, 172n
dāndhā, 153 & n
dar-o-bast, 10, 225
Dar-ul-Khair, Ajmer, 12n
dastūr, *dastūr-ul-'amal*, revenue-rates, 55-9, 84; *dastūrs* on peasants, 124-26; 269, 270-9; fixation of 325-26. See *'amal dastūr*, *dastūr chaudhurī qānūngo*, tax regulation, 48-9, 151-52
dastūr-ul-'amal, 124-26, 269, 270-79, 325-26. See *dastūr* and *'amal dastūr*
Dausa, *pargana*, vills., held by Jai Singh, Sawai, 11n; Expansion of, 26, *tappas* of, 31; revenue-demand in, 56, 97; prices in, 83; sale of grain, 101; *sāir-jihāt*, 150-51 & 157; *Muwāfiq jama'-bandī* & *Siwāi' jama'bandī* 160-61; *pargana* dignitaries, 197-99; vills., in *khāliṣa* and *jāgīr* in, 213; 229; *arhsattas*, 319; *taqsīms*, 323; *jama'bandis*, 324; *nirkh bazār*, 329
deh-ba-dehī, 333
Delhi (Shahjahanabad), 37, 138
Deoli Sanchari, 6, 18, *dahsāla* of, 323
dhān paddy, 173n.
dhenkli, means of water-lift, 148
dīwān, 128-29; appointment, obligation & privileges, 163-66; *dīwān-i des* (*khālīsa*) & *dīwān-i tan*, 164
dol (*daul*), 10n, 172n, 221-22
dorī, unit of length, 41, 290
Durbār (Jaipur), 224, 229-30
ekotri (*ekotra*), 151, 158, 185; see *nirotri*

Fahrist deh-ba-deh, 7n
famines, 139
farah, 145 & n
fard, 27n
farohi quṣur, 158
farmāī'sh, 4n
Farrukh Siyar, 224
faujdār, 3, 13, 15
faujdārī, 14 & n
flour (wheat and barley), 104
food-grains (*jinsī*), 45-50 & 60, production of, 62-3, 74; prices of, 83-6, 92, 100; sale of, 101-02
fotadār (*potdār*), 164-65; pay of, 173; functions and obligations of, 176-77
ganj, 103
ganwars (*gunwars*), 122, 171n

Gaj Singh Kumbhawat, 171n
Gaj Singh Rajawat, 137, 221
garhi, 136, 140n
garhi bhom, 136
Gaonri, *pargana*; revenue statistics of, 40n, 173, 176
gaonveti; nature of, 112, 118, 121, 126
Garh Ranthambhore, *sarkar*, 8; *pargana* and vills., in *jagīr*, 34-7
Gau-shūmarī (*gao-sumari*), 151-52; taxation on, 159
gaz-i-Ilāhi, 41 & n
gaz-i Sikandarī, 280, 290
ghar-baithakī jāgīr, 213
gharoo-hala, 118 & n., 121
gharaicha (*gheruha*), 109, 152
Ghasi Ram, Rai, 15
Ghī (*ghee*); trade in 104n, 107,

111-13; cess upon, 158; prices of, 328
ghiwai (*biyāẓ*), 139. See *byaz-ghiwai*
ghughri shahna, 157, 179
gochani, cereal, 62
gojai, cereal, 62
Gokal Ram Brahman, 175
Gold, 104 & n; prices of, 310-12
Goldsmith, See *Sunar*
Gopal Raipura, 140
goriwa, 146n
gram, pulse; revenue from, 48-50; state demand on, 59, 84; area and produce per *bīgha*, 61-2; prices of 69 & 77, 85; sale and purchase of, 101-02
gūjar, peasant, 129; caste, 131, 135; bullocks possessed by, 131 & 133
Gujrat; trade from, 111-12, 115
gumāshtās, agents of *jāgīrdārs*, 163, 169n, 175
gunjāī'sh, 170, 217
gur; trade in, 113

Habib, Irfan, 41-2, 134, 145, 172n, 218n-19n
habūbāt, 226
hal, plough, 140n, 183
ḥāl-i ḥāṣil (*ḥāl ḥāṣil*); Meaning of, 144; Break-up, 144, 167, 217
Hasan Khan Ghigada Khan, 15
Hasan, S.N., 2n, 44, 74n, 84, 334
ḥasbūl mufaṣal, 144
ḥasbūl waṣūl, 144 & n
ḥāṣil, 139, 152, 156-59; statistics of, 289
hatwāra, 110
ḥavelī, 28, 36-7
Hidayat Kesh Khan, 15
Hindaun, *pargana*, 7; *jama'* of, 19, vills., in *jāgīr*, 32; area of, 43; *bhomiās* of, 135

ijāra, 10, 11, 15-24, 145, 183, 223-32; types of, 223; tenure of, 224-25 & 228; in *khāliṣa*, 226-27; emergence of *thikānās* due to, 230-32
ijārādār, 17, as trader, 103, 142, 169, 224-25; relations with the Jaipur Durbar; 228-32
ijārā istimrārī, 16, 230

Ikhlāṣ Khan, 15
iltamash, 175 & n
Imperial *khāliṣa* and *manṣabdārs*, 8-9. & 9n, 163, 225
in'ām; *jāgīr* assigned as, 7, 11, 165
Indigo; production, 60; prices, 86; trade in, 110-13, 115
Ishwari Singh, Maharaja, 2:0; territories in *jāgīr*, 17 & 25
iẓāfa, 120-21, 146 & n

Jackson, 231
jāgīr, def. 2-4, 6-7; *dar-o-bast* and *shirkat*, 8-9 & n; transfer of 10-12 & 14-5; areas taken from previous assignees, 18-24, 26; held by rulers of Amber/Jaipur, 28-37, 144; sub-assignment and resumption of, 163 & 213-20, 165, 186; administration of, 211-20; revenue from, 289. See *tankhwah jāgīr*
Jāgīrdārs, position and obligation of, 1-5, 14-05 & 135, 17, 25-7, 184, 186; revenue collected by, transfer and resumption of assignments of 218-19; administrative arrangements of, 211; pay and remuneration of, 214-19, 218n
jāgīrdārī, 14-5; month-proportions, 217
Jagjiwan Das, Pancholi, 15
Jahangir, 3

Jai Singh, Mirza Raja; *parganas* held by, 6-7 & n, 12

Jai Singh, Sawai, war of succession, 2; attempt to acquire *parganas* near *waṭan*, 10-11; Expansion of Amber under, 11-27; territories held by, 28-37; Land Revenue Administration, 163-86; list of *dīwān*, *amīn* and *amīn*/Karkun under, 187-210; and evolution of Jaipur state, 211-12; system of *jāgīr* under, 211-22; system of *ijārā* under, 223-30; emergence of *thikānas* and their relations with, 219-20 & 230-32

Jaitpur, *pargana*, 216

Jalalpur, *pargana*,6

Jama', 2, 3, 5-6; statistics of *parganas*; Malarna, Naraina, 9 & 18-24; *pāibāqī* and Imperial *manṣabdārs*, 9n; *jama'* as standard assessment, 216; equal to *tan* in sub-assignment, 216-17

iama'-dol, 70n

jama'-i dahsāla. See muwazana and *taqsīm dah-sāla*

jama'bandī, 92, 102n; sale of grain in different villages in documents of, 101-02; statistics of, 244-54

jarīb, 280

jarībāna; rate & collection of, 145, 151

Jaswant Singh, 221

Jāts, *waṭan* of, 8; holdings of. 24; peasant & castes of, 67-8 & 135; *asāmīs* and bullocks held by, 131-32; revolts of, 137n, 138 & n

Jat Kalyawa, *pargana*, 6 & n

Jhak, *pargana*, 6, 34

Jhilai, *pargana*, 8

Jhunjhunu, *pargana*, 135

jumla (*jumlo*), 128

jhuprī, cess, 151, 157

jihāt, 145-46, 148, 280; rate of, 124 & 126

jinsī, 39-40, 44, 50, 60, 62; share in revenue of, 64-6, 92-3; sale of, 101-02; 145

jinsī system, 40, 43, 100, 114, 126, 140

Jodhpur, 3

jōt, 42-3

Jowār (*jowari*), millet, revenue from, 45-8; state share in, 59; area under, 61; prices of, 80

juro, 137

Jute, 110

Kachhwahas, 163, 186, 211, 224; position in the Mughal Empire, 1-5; expansion of territories of, 5-37

Kaghazāt-i Mutafarr iqa, 32, 35

Kaladera, *pargana*, 26

Kalāl, 109

Kalibhit, 121

Kalyan Das, 93, 121, 138

Kalyanots (Kilanot), 135

Kamīn, tax on, 124-26, 285

Kārkun, 167, 173. See *'āmil*

Kayāl (weighman), 107

Kayālī, cess, 107 ; meaning of, 107; revenue realised under, 151; rate of, 157

Kayastha, Kayath, 212

Kesari Singh, 212

Keso Das, 121

Khāliṣa, *Khāliṣa sharīfa*, *Khāliṣa Pādshāhī*; vills., in, 8-11, 212-13; in *jāgīr*, 39; of Amber ruler, agricultural production, 74-91; taxation, 144-62 & 186; administration, 163-86; 212; revenue in, 213-14; revenue-grants in, 211-16; revenue-farming in, 226-29

Khānd, 111-12

Khandela, *pargana*, 173, 220

Khangarots, clan, 135

Khārwal, caste, 131; tax upon 152n
Kharīf, 45, 59; *kharīf* crops, 60-01; 64-6, 74, 84-5
Khasra (*khasro*, *kasro*), 127-30, 327-8; of vill. Morara, 254-309; Naraina, 294-309
Khāta, 102
Khāti, 132
Khatīk, 109
Khatrīs, caste, 174
Khetri, *pargana*, 135
Khohri, *pargana*, 6, 18-9, 24, 26, 112, 228n
Khoonchi, 107
Khud-Kāshta, 118 & n, 119, 122, 133
Khūnch, 108
Khurāk, *khurakghora*, 219
Kirāna, 111
Kirpa Ram, 222
Kishtikār, 42n
Kodo, *kodron*, millet; revenue from, 45-8, 55-9; area under, 61
Kol, *pargana*, 33, 138n, 325
Kori bhomī ki, 106
Kota, *pargana*, 130
Kotla, *pargana*, 101-02
Kotwālī, cess, 106
Kumbhār, 67-8, 131, 300
Kūt, 177.
Kyārī, 70 & n, 71

Labourers, 122; hired labour, 118
Lalsot, *pargana*; vills., 8, 32; revenue in, 23, 98, 135, 148; revenue from, 57; prices of food-grains in, 83; *sāi'r-jihāt* in, 151, 154; *muwāfiq-jama'bandī* and *siwāi' jama'bandī* of, 161, 162, 173, 229
Lakhtāng, 137, 140, 213, 228n, 232
Land-revenue, 92; realization and burden of, 124-26; magnitude of, 153-55; assessment of, 144-55; farming of, 223-30; mode of payment of, 19; collection of, 163-77; remissions and concessions in, 59, 124-26; assignment of, 213-16
Land rights, 116-33
lātā (lato, latai), 177; method of assessment, 177n—78 & 178n
Lawan, *pargana*, 7
lawāzima (*lawāzimo*), 145 & n
lāyaq-i zarāi't (*layaq zaraitī*), 41, 116, 291, 293
Liwali, *pargana*, 101
ma'āfī, 219
madad-i ma'āsh, 171

Madho Singh, 25
magra (*magro*), 41, 146n
Mahājan, grain-dealer, 93, 100, 105; as *ijārādār*, 106 & 227; favoured revenue-payer and privileged class, 123 & 124-26, 127, 130; position of, 140-43, 164n, 174, 176-77; as *pargana* official, 206-10; as *jāgīrdār*, 212
mahal, 13n, 29, 166, 225
Maharaj Kumar, 138
Mahindār, 177-78
majurs, 121
Māl, 126, 145. See Land Revenue
Malarna, *pargana jama'* of, 8-9 & 21, vills. in *jāgīrs*, 35, 38-9; cropping in, 42 & n 43 ; revenue from, 45-9, 55-56; revenue from wheat, 54; State share, 59; area under different crops in, 61-2; prices of food-grains in, 83-5; revenue in cash and kind, 99; sale of grain in, 102, 116 & n, 117 & n, 135; *sāir'-jihāt* from, 150-51, 157; *muwāfiq-jama'bandī* and *siwāi'-jama'bandī*, 161-62; pay of *pargana* officials of, 173; '*āmils*, *amīns* and other officials, 174 & 193-96; Maharaja's *Khālisa* in, 213-229
mālī, caste, 67-8, 133
Maka (maize); revenue from, 45-8, area under, 61, 62
malba, 153 & n., 182

māl-zāminī, 176, 229
māl-o jihāt, 145, 153
Malwa, *sūba*, 11, 22, 121
man, unit of weight, 39, 74; *sers* in, 75 & n
Mandhwah, millet, 40n, 60
manṣab, 6, 12
manṣabdār, 4, 9n, 13, 15, 92, 225
Man Singh, Raja, 3, 6
Manoharpur, *pargana*, 11, 220
Māpa, cess, 106, 156-57
Marathas, 17, 211, 230
Marwar, 134n
Mārwārī, 130
Mathura, (Islamabad), *pargana*, 32; *sarkār*, 33
Maujpur, *pargana*, 26, 29
Mauzabad, *pargana*, 6, 9, 12, 67-8, 93, 101, 131
Mercantile classes, 115, 140-43
Merchants, urban, 105, 108-13
Mewat, 1, 16
Minas, 135, 181
Minzālik, 167
Mirza Raja, 7, 12
Mirzas, 5
Money-lenders, 105, 119, 140
Month-proportion or Month ratios, 217, 218 & n
Moth, pulse, 40n; revenue from, 45-8; area under, 61; prices of, 81, 83, 85; sale of, 101
muchalka, 7n, 25n, 180n, 215, 333
Mufti, 103
Mughal Empire, 1-5, 10, 14-7, 42, 92, 134, 155, 163, 174, 186, 218-19, 224-25; administration, 217; revenue system, 155; territory, 211
Muhammadabad alias Ghazi ka Thana, *pargana*, 15, 18, 23, 31
muhr, 310-12
mujmil, 167n, 318
mujrāi, 183
Mukhia, 180
Mundawar, *pargana*, 102
mung, mash, pulse; area and production, 40n, 61, 74; revenue from, 45-8, 55-9; state share in, 59; prices of, 83, 85; standard demand per *bīgha*, 84; sale of, 101
muqaddam, 180
muqarara jama', 167, 176
muqta'ī, 176
murādī, copper coin, 74n
muṣā'adat, 15, 175n, 219
mustaufī, 166 & n
mutāliba, 219 & n
muwāfiq jama'bandī, 145, 153; size of, 160-62
Muwāzana, 27 & n
Muwāzana-i dohsāla, 41n, 323
nābūd, 41 & n, 290-92

Nakhkhās, 107
'nalbant, 153 & n
Nanig Ram, 166
Nānkār, 42n, 176, 183 & n
naqdī, 126, 215, 220n
Naraina, *pargana*; its *jama'*, 9 vills. in *jāgīrs*, 37; measured area, 44; total taxation in, 154-55; pay of official in, 173
Narnaul, Narnol, *pargana*, 15, vills. in *jāgīrs*, 29
Narukas, 135
nazrāna, 4n
Nihar, *pargana*, 7
Nirotri, 158
nirkh bāzār, 328-30
Nirwan, 212
Nishān, 6n
Niwai, *pargana*, 7-8; vills. in *jāgīrs*, 35, 41n; measured area, 43, 322-23
pachduvo, 59, 129
Pachwara, 6; *parganas* in, 6n & 32

pā'i (*pāhī*), 118, 119 & n, 124-25
pā'ikāsht (*pāhikasht*) 118; rights of 118-19 & 119n; concessional rates of tax upon, 125
pā'ibaqī, 9 & 9n, 12n, 13n
paisa, copper coin, 172 & n
pāltī, *palatiyas*, *paltaya*, 120-21; origin and occupancy rights of, 122; tax on 124, 133, 141, 148, 171
Panch, 169n
Panchanot, clan, 135
Panchpana-Singhana, *pargana*, 231
Panchāyat, *panchaiti*, 127-29
Panch-mahājan, 141n
pāni, 156
Panjab, 6
Paparda, *pargana*, 7
parat, *parautī*, 43, 119, 170n
pargana functionaries, 166-74, 187-210
parganas; *jama'* of, 8-9, 18-24, 293; vills. in *jāgīrs*, 11n & 12n, 28-37; held by Amber rulers, 28-37; officials & functionaries of, 163-210
Patan, *pargana*, 26
paṭel, *khud-kāsht* peasants, 102, 121; favourcd revenue-payers, 123-27; tax on, 124-26 & 128-30; bullocks held by, 67-8 & 131-33, 146, 149, 165; obligations and privileges of, 180-84; castes of, 181, 227
paṭeli, 181n
paṭeli ka khera ka dol, 183
paṭṭa (s), document, 119, 168, 171n, 223, 225, 228
paṭṭīs, 127, 129-130
Paṭwāra, *patwaro*, cess, 151, 185
Paṭwārī, 102, 121, 131, 164-65, 168; position, functions, perquisites and income of, 184-86, 228
Peasants, bullocks held by, 67-8, 131.3; trade by, 101-03, 112; stratification, 118-31; land-revenue from, 124-26; oppression of, 135-36, 139, 141, 143, 155, 227
Pelsaert, 115n
peshkash, nature of, 3 & 152, 169, 181 & n, 214
Phagi, *pargana*, 6-8, 101, 135
Pilibhit, 121
Piragpur, *pargana*, 121
piyādgī, 153n
pīwal, *pīyāl*, 148
Ploughs, see hal
polaj (*polach*), 59, 125
Ponotra, 142, 158
Poṭah-khāna, treasury, 166
Poṭdār, 164-65; pay of, 173; functions, 177
Prices, Agrarian, 74-90, 323-30; copper, silver and gold, 310-16
Pulses, 63
punya-udik, 42, 144, 165, 171, 233
qabūliyat, 223, 227n

Qaimkhanis, 231
Qānūngo, concessional tax upon, 124-27, 146, 148, 164-65, 167-68, 170; hereditary office of, 174; functions of, 175-76; perquisites of, 176
qarār-ijāra, 45, 145
Qāzī, 228
rabī', *rabī'* crops (sakh unahlu), revenue from 48-50, 59, 64-66, 69-73; area under, 61-2; principal crops of, 62-3; prices of, 75-7; trade in, 101-02
r'ahdārī, 103, 106; *dastūr rāhdārī*, 110-12
ra'iyat (peasant); and trader, 102 & 115, 118; tax upon, 123-25, 153-55, 166, 169, 171, 176; classes of, 139, 142
ra'iyatī villages, classification of, 116-17; tax regulations of, 117

ra'iyatī-kāsht, 118 & n
Rāj, 163 & n, 164, 188n
Rāja (s), title granted by the Mughals, 1; position of, 1-5
Rājāwat, clan, 1, 4, 135, 137, 212
Rājāmahal (Akbarnagar), 6
Rajputana, 17
Rajputs, *zamīndārs*, 1, 4, 6, 117, 120, 124-27, 130, 135, 143, 146, 212, 225
Ram Singh, Maharaja; *parganas* held in *J.*, 7 & n, 22, 232
Rampura, 30
Ranthambhore, *sarkār*, 34-6
Revenue, 92-3, 115, 117, 144; from different crops, 45-54; demand per *bīgha*, 55-9, 88; see *ẓabṭ*.
rozīna, 157-58
Roznāma, *Roznāmcha*, 332
Roznāma-wāqi a, 93n
raqba, *raqbo*, 41, 233, 236, 290, 292

Sahar, *pargana*, 33
Sahardār (Shahardār), 105
Sāh, 174, 188-210
Sahthānī, 105
Sahūkār, 15, 141, 180, 229
sāir, *sāi'r jihāt*, 145; income from, 150; various items under, 151, 156-59, 171
salt, 111-12; trade in, 111-12, 122n
Sambhar, *pargana*, 14n, 112
Sanganer, *qaṣba*, 1
sanjhādār, share-cropper, 130
sārh-athotri, 158
sarhi, 158
Sarkār, 8-9, *parganas* held in, 28 & 33-4, 36-7, 124-26
sarso, 62, 101-02
Sawai, Jai Singh; See Jai Singh, Sawai
Sawai Jaipur, 1, 14; founded, 36; revenue from, 45 & 48-9, 51, 55, 64; under different crops in, 61-2; land-revenue administration of, 163-86; officials of, 180-210; *jāgīr* system of, 211-22; sovereignty of rulers of, 219-20; *ijarādārs/thikānādārs*, 230-31. See also Amber
sar-talab, 172n
sawār, 2, 6, 214
sarwālo man, 39 & n, 248-49
Sekhawat Confederation, 10n, 231
ser, unit of weight, 75 & n
Shahjahanabad (Delhi), *ṣūba*, 37
Shahna, 151, 153, 157; appointment of, 178; functions and perquisites of, 179
Shahpura, *pargana*, 26
Sharfuddin Mirza, 1
Shekhawat, clan, 4; 127, 146, 231-32
shirkat (*sarikat*), 8, 215, 226
Silver, 89, 104 & n, 176, gold ratio, 310-12; copper in relation of, 313-16
singhāra, 101
Singoti, 107
sīr, 42 & n., 216; area under, 289
Siwāi'-Jama'bandī, 144 & n; income from items of, 152-53; nature of items, 152-53; size, 160-62
Sobner, Jobner, *pargana*, 8
Stratification, vill. society, 116-33
Ṣūba, 8-10; 28, 34, 37
Sugar, trade in, 110-13; prices of, 328
Sugarcane, *varh*; revenue from, 45-8, 55-60; area under, 61
sunār, Goldsmith, 67, 301
Siyāqnāmā, 218 & n
Syalu, *kharīf*, 169n, 184
tābīnān, 2
ta'alluqa, 116, 117 & n
ta'alluqadārs, 117, 176
tafāwat, 152
taḥṣīldār, 101, 176

taḥṣīldārī, 151, rate of, 158
taka, copper coin, 108, 313-14
takhfīf, 170, 217, 228
takhmīna, 40n, 41, 69n, 168n, 176; nature of, 323
ṭalab, meaning of, 152 & n, 217-18
ṭalbāna, 152 & n
tan, 216, 217
tappa, 28-37, 166 & n, 233
tappadār, 167, 177-78; pay, 178 & n
tappadārī, *rozīna tappadārī*, 151-52
taqāvī (*taqai*, *takai*), 142, 167, 182
taqsīm, papers, 3n, 43, 175, nature of, 289-93, 322-23

Taxation, 108-09, 123, 127, 144-62; burden of agricultural, 153-55; under *sā'ir-jihāt*, 156-59; under *siwāi'-jama'bandī*, 152-53
Tazikra, (*tazkara*), 127
tehbāzārī, nature of, 107 & 153 & n
telī, bullocks held by, 67-8, 100, 102, 131, 133
thikāna, 135
thikānādārs, 220 & n; emergence of, 230-32
Tijara, 15
til, oilseed; area & production under, 40n; tax on, 59, 101-02; trade in, 110 & 112-13
tīka, 3
Timur, 5
Tobacco; cultivation of, 60, 62; trade in, 110-11
Tod, James, 10n, 134n, 135n, 138n, 231
Toda Bhim, *pargana*, 33, 225
Tonk, pargana; tax on *asāmī* of, 128, 135
Trade in, 92-115, 104; transit duties, 108, 113
tulai, cess, 151, 157
Turks, Turak, 13
Udai, *pargana*, vills. in, 31; measured area in, 43, 139n
ujār, 67, 185

Umran, *pargana*, 6
Unahlu, *rabi'*, 184
Uniara, *pargana*, 135
Urban merchants, 105

Vani, cotton, 146n
vakīl, 10n, 15 & n
Vārhdār, 41n
vasıdār (s), 136
vāt, 59
vātvīh, *bātvīh*, 119, 121, 153
vegetables, 60, 104, 147
Vir Vinod, 30
Village, *mauza*, in *jāgīr*, 28-37, 39; actual cultivated area in, 42; peasant and bullocks in, 63 & 67-8; sale of grain and trade in, 101-02 & 105 & 108; division of, 116-17; stratification in, 118-33
Village community, 127
Village officials, 180-85
vor, 70n, 148

Wage-labours (*majurs*), 121
Waqia-navīs, 152
waṭan, nature of, 1, 6, 7, 10n; expansion of, 11, 13, 164
waṭan jāgīr, 2, 10n
Wheat, 40n; revenue from, 48-50, 51-4, 59; area under, 61-2; cultivation of, 69-73; prices of, 76, 85; trade and transport of, 101
yāddāsht, *yaddāsti*, 103n, 105n
Wills, C.U., 12n, 18; inquiry, 220, 230-31

Ẓabṭ, ẓabṭī, 39 & n; rates under, 44 & 50 & 55-59 & 146; crops under, 62

Zamīndārs, 1; vassal chief, 1 & 2; as a class, 117, 127; land revenue paid by, 124-26; rights of, 135

Zamīndārī rights, 10n, 177; hereditary, 181; caste possession, 131-33; *zorṭalab* (rebellious), 169 & n